AF376210

6 Crystals

Growth, Properties, and Applications

Managing Editor: H. C. Freyhardt

Editors: T. Arizumi, W. Bardsley, H. Bethge
A. A. Chernov, H. C. Freyhardt, J. Grabmaier
S. Haussühl, R. Hoppe, R. Kern, R. A. Laudise
R. Nitsche, A. Rabenau, W. B. White
A. F. Witt, F. W. Young, Jr.

Field-Ion Microscopy

By R. Wagner

Springer-Verlag
Berlin Heidelberg New York 1982

ISBN-13:978-3-642-68689-4   e-ISBN-13:978-3-642-68687-0
DOI: 10.1007/978-3-642-68687-0

Library of Congress Cataloging in Publication Data. Wagner, R. (Richard), 1947–. Field-ion microscopy.
(Crystals-growth, properties, and applications; 6) Bibliography: p. 1. Materials – Microscopy. 2. Field ion
microscope. I. Title. II. Series. TA417.23.W33 1982 502′.8′2 82-10782 ISBN-13:978-3-642-68689-4(U.S.)

Table of Contents

Field-Ion Microscopy in Materials Science

Richard Wagner

GKSS-Research Centre, Geesthacht, FRG, and Sonderforschungsbereich 126, Göttingen-Clausthal*

During the past decade analytical field-ion microscopy comprising field-ion microscopy (FIM) in combination with single-atom mass-spectroscopy (atom-probe techniques) has been developed to become a powerful microanalytical tool in physical metallurgy, particularly suited to the detection of ultra-fine defects in metals and alloys, and to the analysis of their chemistry. In this article we have attempted to cover all aspects and recent developments in analytical field-ion microscopy which are relevant to the application of this technique to problems in physical metallurgy. Emphasis is made on the discussion of various methods for the quantitative analysis of short-ranged or long-ranged composition fluctuations such as those resulting from various phase transformations in alloys, and also on the discussion of factors which govern the spatial resolution of the atom-probe. Insight into the range of applicability of the various field-ion-microscope techniques is gained from a comprehensive discussion of recent results from investigations utilizing these techniques to study point defect-configurations, the structure and topography of interfaces, and various kinds of phase transformations in metals and alloys, and also to investigate metallic glasses and semiconductors.

* Habilitationsschrift, Univ. Göttingen

1 Introduction

Despite the recent progress in developing various microanalytical tools of better spatial resolution and more sensitivity to chemical analyses for the study of various defects in metallic solids the Field-Ion Microscope (FIM) still remains the only instrument up to now to resolve single atoms in the surface of a metal. Fifteen years after Müller[1] invented the FIM he was also the first to combine the FIM with a time-of-flight (ToF) mass spectrometer – the so-called *Atom-Probe FIM* – to identify the chemical nature of single atoms imaged in the FIM[2].

Originally the motivation to develop the ToF atom probe was to use this method to obtain some more fundamental understanding of field ionization and field evaporation, the most basic physical processes in field-ion microscopy. Even after the successful combination of a FIM with a ToF atom probe had been accomplished, the technique was rarely applied to metallurgical investigations since for a fairly long period only refractory metals such as tungsten, molybdenum, iridium, etc. could be imaged in the FIM. However, these metals do not play a very important role in metallurgy. Only when Turner et al.[3] substituted the conventional phosphorescent screen of the field-ion microscope with micro-channel electron multiplier arrays, termed microchannel plates, did it become possible to image in the FIM the less refractory metals like Fe, Cu, Ni and even Al. These metals and their alloys are more relevant to materials science and, therefore, more interesting for basic investigations of the defect structure in these materials. By making use of the controlled field-evaporation technique (Chap. 2.4), the analysis of the material under investigation by means of the atom probe FIM is no longer confined to the topmost surface layer imaged in the FIM, but rather allows the three dimensional distribution of the defects in the bulk material as well as the topology of a particular defect to be investigated, with a depth resolution of approximately 0.2 nm. It is this extremely good depth resolution together with the unique combination of atomic resolution and single-atom mass spectroscopy which makes the ToF atom probe FIM so well suited to the investigation of defects having dimensions of one atom up to approximately 25 nm.

As will become obvious from the description of the principle of the atom probe (Chap. 3.1), the area analysed with the atom probe is limited to the area of the FIM image covered by the aperture of the atom probe. This aperture, termed probe-hole, usually has a diameter variable between one and a few nm in a standard atom-probe FIM designed for metallurgical applications. This limited probe-hole size offers a good lateral resolution which is often advantageous for the chemical analysis of small precipitates (Chap. 3.2) or the analysis of one-dimensional concentration fluctuations of short wavelengths. However, a considerable amount of information, which is contained in the surface of a field-ion tip, is lost by this probe-hole technique. This is a disadvantage for some metallurgical applications but has been overcome quite recently by the development of the *Imaging Atom Probe (IAP)*[4] (Chap. 3.5), originally termed field-desorption mass spectroscopy by Panitz[5, 6], who first was able to handle the experimental difficulties associated with the IAP. For some metallurgical applications, e.g. detection of sharp concentration gradients or grain boundary segregation (Chap. 4.5.3) the IAP shows some advantages compared with the conventional ToF atom probe. Today the experimentalist has to consider his specific metallurgical problem to decide which of the two field-ion spectroscopic techniques, conventional ToF atom probe (ToF-AP) or imaging atom probe (IAP), will yield more quantitative information about the defect structure of his metal or alloy. Since both techniques are based on the same physical processes, it was only an instrumental difficulty to combine a ToF-AP with an IAP into one FIM system[7, 8] to make use of either technique in various metallurgical investigations.

It is the purpose of this article to review the recent applications of quantitative field-ion spectroscopy to problems in materials science. For this purpose the rather complex basic physical processes, as FIM image formation and field evaporation, are only described to such an extent as is necessary for the understanding of the specific problems which frequently arise during imaging defects in metals and alloys, or when analysing quantitatively the composition of either single-phase or multi-phase alloys. Detailed discussions of the mechanisms of field ionisation and field evaporation may be found in the books of Müller and Tsong[9] and Bowkett and Smith[10].

The principles and applications of the FIM without ToF atom probe to investigations of defects in mainly pure metals like W and Ir, which had been performed prior to 1969, have been presented in a detailed comprehensive review by Bowkett and Smith[10]; some results from studies of grain boundary structures and especially from studies of the core structure of dislocations by means of FIM are also included. Since then, because of the availability of the atom probe FIM, most work has concentrated on studies of phase transformations such as the early stages of precipitation reactions (Chap. 4.6), order-disorder reactions (Chap. 4.4) and on studies of radiation induced defects in metals and alloys (Chap. 4.3). Also the type of materials investigated has changed. Nowadays the quantitative field-ion spectroscopy is applied to investigate phase transformations in technically important alloys such as steels or Ni-based superalloys (Chaps. 4.5.3, 4.6.2). During the last two years some progress has even been made in imaging and analyzing semiconductors (Chap. 4.7) and amorphous metallic glasses (Chap. 4.8).

2 Principles of FIM Technique

2.1 Specimen Preparation

The typical FIM specimen is a sharply pointed needle with a radius of curvature ranging between 20 and 80 nm. The starting material for preparation of a FIM tip is usually a thin wire of 0.1 to 0.3 mm diameter, which is either obtained by cold drawing or by spark machining from a macroscopic piece of material. Prior to anodic electropolishing of the wire to form a sharp tip, all the thermodynamic heat treatments necessary for the particular metallurgical study have to be performed, e.g. annealing of the wire to remove the lattice defects and distortions introduced by the cold drawing, or solution and aging treatments to initiate the wanted phase transformation. This latter point is of special importance, since sometimes the course of a phase transformation might be altered in the case of an *in-situ* heat treatment of the already sharpened FIM tip where the ratio of the number of atoms in the surface to that in the bulk is fairly large.

The actual FIM tip is prepared by immersing the wire vertically to a depth of 10 to 15 mm into an electrolyte (Fig. 2.1). The electropolishing of the wire is done with AC or DC. The taper angle which forms at the end of the dipped wire is controlled with an optical microscope; the deeper the wire is dipped into the electrolyte, the smaller the final cone angle of the tip will be.

There have been proposed several polishing techniques and different electrolytes for various materials[10, 11]. However, based on our own experiences, we generally found the method shown in Fig. 2.1 to be suited for all materials; we further found that most electrolytes and polishing conditions which are recommended to prepare thin foils from different materials for transmission electron microscopy (TEM) (e.g. Ref. 12) are also suited to prepare FIM tips. Sometimes it is useful to investigate the shape of the polished

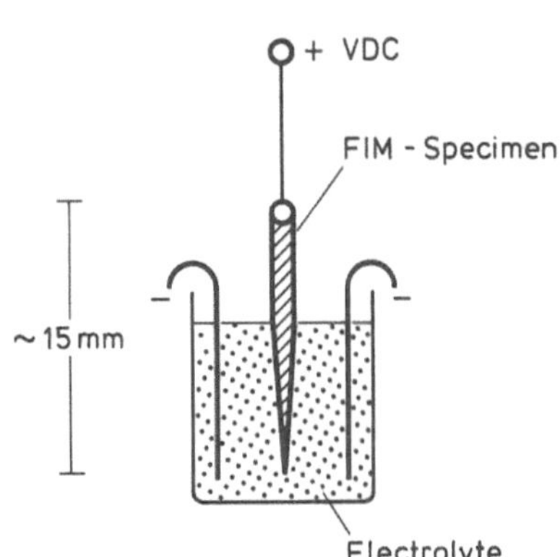

Fig. 2.1. Preparation of a field-ion tip by electropolishing. Usually the tip formation is controlled through a microscope of long focus

FIM tip in the TEM or scanning electron microscope (SEM) (Fig. 2.2); for this purpose special TEM specimen holders for different electron microscopes have been designed[10, 13]. The use of TEM even becomes unavoidable if one wants to have purposely a grain boundary cutting the surface of a FIM tip, e.g. for studies of grain boundary structures and/or grain boundary segregation (Chap. 4.5). Then first a tip is prepared and examined in the TEM for the location of a grain boundary with respect to the tip apex. Afterwards the tip has to be backpolished in a controlled manner by knowing the polishing rate until the grain boundary cuts the tip surface[14].

Sometimes it is not possible to prepare wires either by drawing or by spark erosion. This is the case for brittle semiconductors (Chap. 4.7) and for metallic glasses (Chap. 4.8). The latter are usually only available as thin ribbons (with a cross section of approximately 0.05×1.5 mm^2). This geometry is not suited to prepare FIM tips with rotational symmetry and, therefore, will lead to pronounced streakings in the FIM image due to the strong asymmetry. To obtain symmetrical tips from the metallic glass ribbons, it is necessary to prepare strips of square cross section (0.05×0.05 mm^2) by photo etching[15]. The preparation of FIM tips from semiconductors can be carried out using crystal-growth techniques. Si and GaAs show a crystallographic selectivity of the chemical polishing solution resulting in a blade shaped specimen rather than in a conical tip. This problem may be avoided by using Si whiskers with $\langle 111 \rangle$ major axes as starting material. After dipping the whiskers into a solution of nitric acid (60 parts) and hydrofluoric acid (40 parts) sharp tips could be obtained[16, 17].

After the tip has been polished chemically or electrolytically it looks rather smoothly curved even on a scale which is conveniently resolved by SEM or TEM (Fig. 2.2); however, on a even finer scale which is only resolved in the FIM the tip surface looks rather rough.

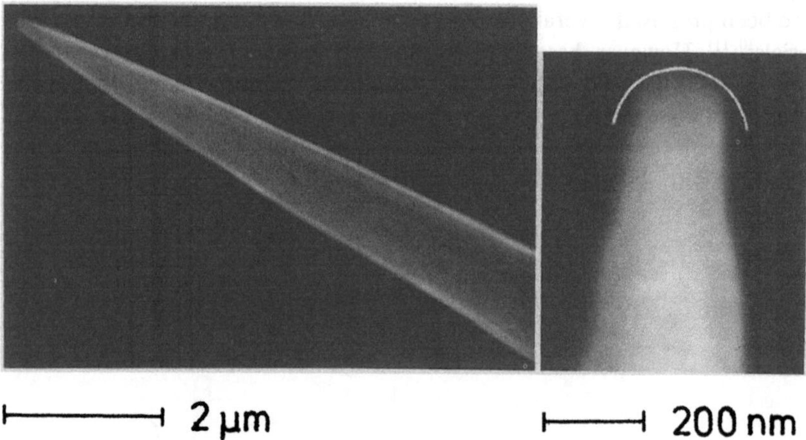

Fig. 2.2. SEM micrographs of typical field-ion tips (iron) which could be imaged in neon at about 10 keV

2.2 Operation of the Field Ion Microscope

Once the tip is prepared it is introduced into the FIM via a specimen airlock (Fig. 2.3). The FIM itself consists of a vacuum chamber with a background vacuum better than 10^{-8} mbar, an imaging screen composed of a channel plate electron multiplier and a fluorescent screen. The field ion microscopes designed for metallurgical applications differ from those originally designed for surface studies in the following two points: a) the specimen is mounted onto a manipulator which allows the specimen to be rotated independently around two orthogonal axes and to be aligned along the axis of the field ion microscope[18]; the centre of rotation then lies in the microscope axis and is identical with the tip vertex. b) The distance R between tip vertex and center of the screen can be varied between approximately 40 mm and 180 mm (all the technical data given may slightly vary from design to design; the data given here refer to the system built at the University of Göttingen[7, 8].

To obtain a FIM image an imaging gas – for most non-refractory alloys to which we refer in this article it is neon, whereas for refractory metals, helium is used commonly – is leaked into the vacuum system to a pressure up to 5×10^{-6} mbar. Subsequently a positive high voltage is applied to the specimen, which typically ranges between 3 and 15 kV depending on both the imaging gas and the radius of curvature of the field-ion tip. If the resulting electric field exceeds a critical value (about 20 to 50 V/nm) the imaging gas atoms are positively ionized in the high field regions above the tip surface and will be accelerated along almost radial trajectories towards the grounded FIM imaging screen which lights up at the points of impact.

Starting with a freshly prepared tip, the electric field is highest above the sharp protrusions which are retained from the specimen preparation. As the field is gradually increased these protrusions will be removed first by field evaporation (see Chap. 2.4) and with further increasing voltage an atomically smooth curved tip-end form is finally obtained. In this state, the smallest electric field sufficient to ionize the imaging gas atoms is reached only above atoms protruding from the surface. In pure metals these protruding atoms mainly correspond to atoms located in ledge- or kink-site positions of the edges of different stacks of lattice planes (Fig. 2.4 a). The edges belonging to one stack of {hkl} planes form almost concentric polygons or rings because of the intersection of each stack with the more or less spherically shaped tip surface (Fig. 2.4 b). This arrangement of atoms surrounded with high-field regions leads to the well known crystallographic pattern of field ion micrographs of pure metals (Fig. 2.5).

2.3 Image Formation

In the high-field region of the tip, which is cooled down to temperatures below approximately 80 K, the image gas atoms become polarized and are therefore attracted to the positively charged tip surface. Due to this dipole attraction, the polarized gas atoms impinge on the cooled tip surface with a kinetic energy of the order of 0.15 eV under typical FIM imaging conditions[19]. During collision with the tip surface the gas atoms lose

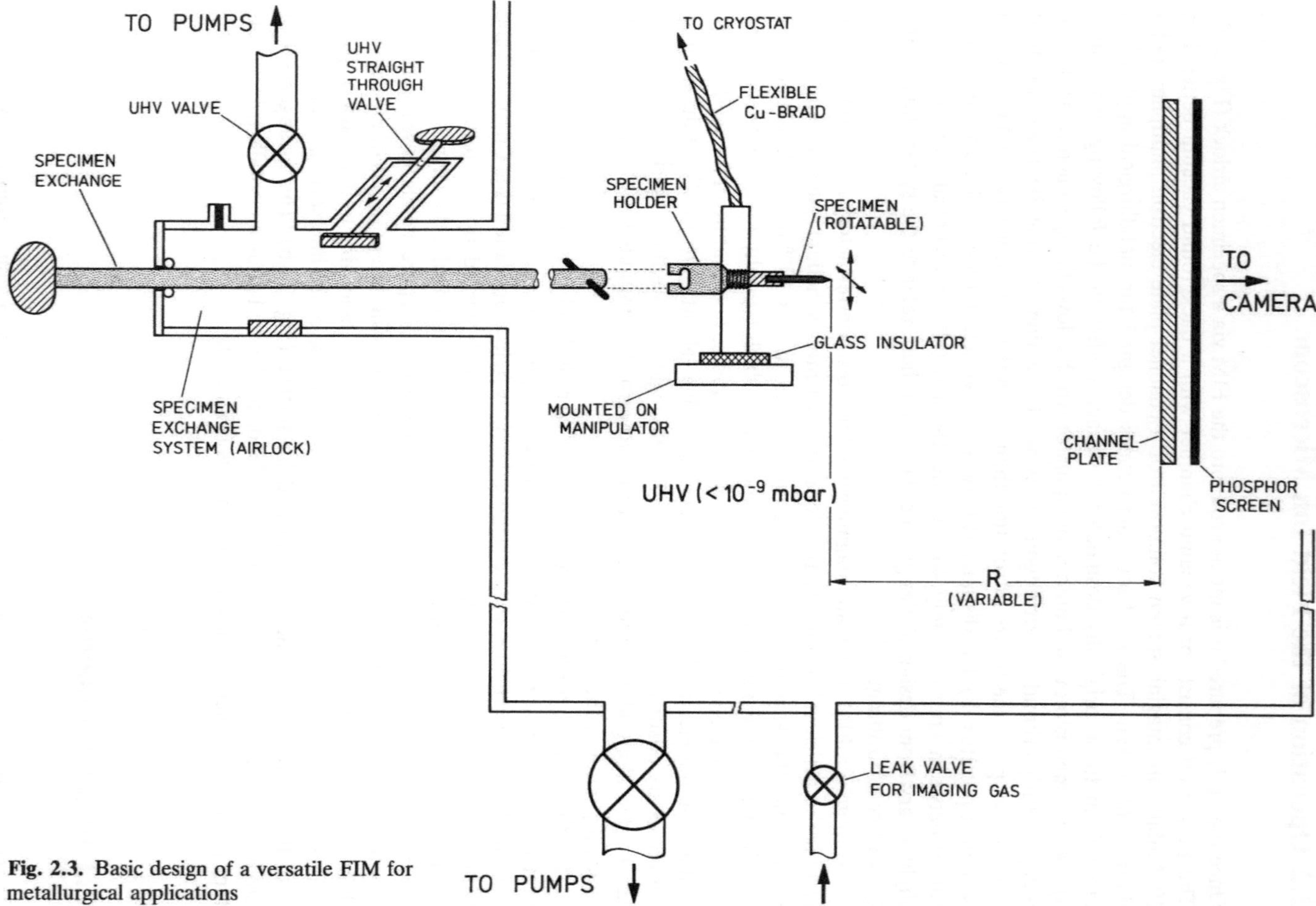

Fig. 2.3. Basic design of a versatile FIM for metallurgical applications

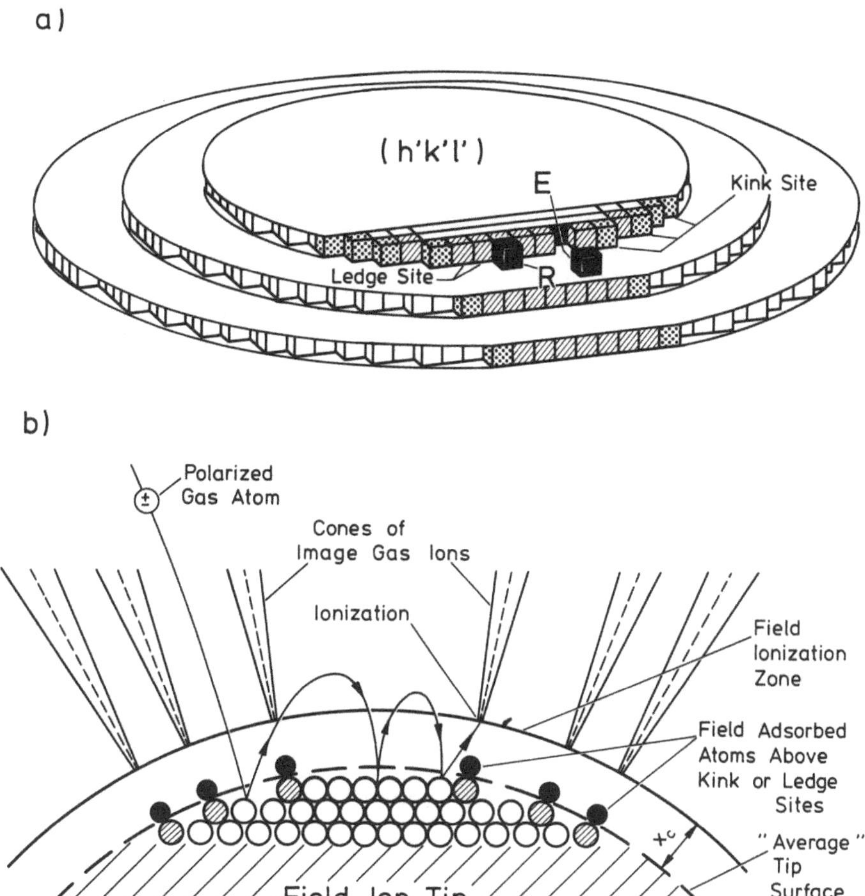

Fig. 2.4a. Three successive layers (h'k'l') of a field-ion tip. The top layer shows ledge-site and kink-site atoms as well as two retained atoms (E, R) in low coordination sites. **b)** Principle of image formation; only the hatched atoms are imaged

part of their kinetic energy to the lattice and, provided the energy loss has been sufficient, the polarized gas atoms are trapped in the high field region. Here they move over the tip surface by ballistic jumps. During each impact with the tip surface part of the kinetic energy is dissipated resulting in a decreasing jump height (Fig. 2.4b) until the kinetic energy of the hopping atom is accommodated to some extent to the low tip temperature[20]. Ionization of the hopping gas atoms occurs only in a very narrow ionization zone at a distance x_c above the tip surface by tunneling of an electron from the gas atom through a potential barrier into the empty electron levels of the metallic tip[19]. The potential energy diagram of an outer electron of a gas atom in the high electric field near the tip surface is illustrated in Fig. 2.6. Close to the tip surface the height of the potential barrier is reduced because of the mutual attraction of the electron and its image charge induced in the metallic tip surface. The closer the atom gets to the surface, the smaller

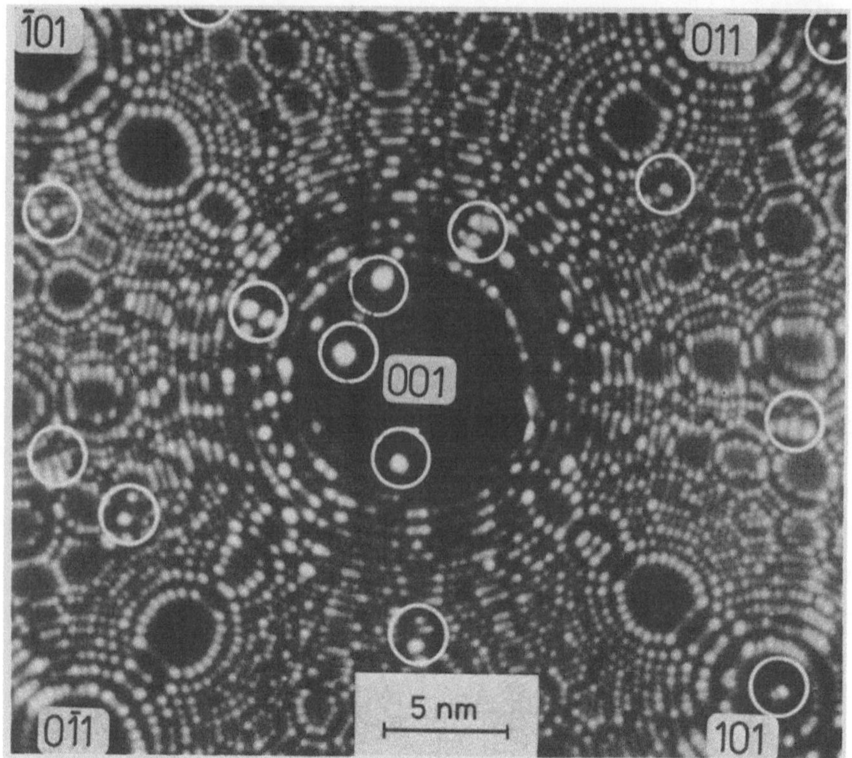

Fig. 2.5. Neon field-ion image of Cu-1.7 at% Fe aged for 1 h at 500 °C. Fe-rich clusters are shown in enhanced bright contrast (*white circles*)

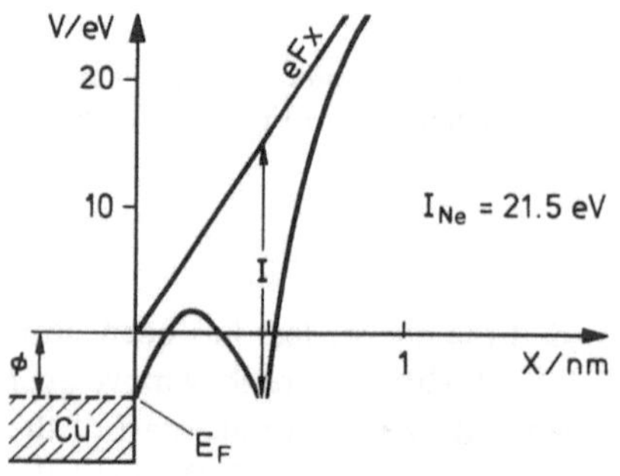

Fig. 2.6. Potential-energy diagram for an outer electron of a neon atom in an electric field near the surface of copper. E_F denotes the Fermi level

the barrier height becomes, associated with an increased electron tunnel probability. However, at distances x smaller than a critical distance x_c, the energy of the electron lies below the Fermi energy of the metal where only few empty electron states are available. Therefore, the tunnel probability is almost negligible for $x < x_c$ and field ionization only occurs in a very narrow region around x_c. To a good approximation[19], the critical distance for field ionization, x_c, is given by

$$x_c = \frac{I - \phi}{eF} \qquad (2.1)$$

where I is the ionization energy of the image gas atom (Ne : 21.5 eV), ϕ is the work function of the metal and F the (local) electric field above the metal atom to be imaged. In the case of copper imaged with neon, Eq. (2.1) yields $x_c \approx 0.49$ nm for a field strength F = 34.5 V/nm. The latter field strength – the so-called *best image field*[21] for neon as imaging gas – yields a maximum of details in the overall field ion image. The optimum field strength to image a particular region of the surface may differ considerably from the overall best image field because of the local variations of the radii of curvature. This holds especially true in the case of two-phase alloys (Chap. 2.6.3).

So far, in considering FIM image formation, it has been assumed that the surface of a field ion tip is completely clean. However, theoretical considerations[22, 23] supported by atom probe studies of field evaporated surface atoms[24, 25] and by studies with the field desorption microscope[26] have established that all brightly imaged atoms in the edge of a {hkl} lattice plane are capped by a field adsorbed image gas atom (Fig. 2.4b). The necessary binding energy between the protruding metal atoms and the gas atom, which is in the order of 0.14 eV under best image field conditions[27], results from the complex dipole-dipole interaction of the polarized gas atom and the substantially positively charged metal surface[23]. The presence of field adsorbed gas atoms above protruding atoms leads to the conclusion that field ionization takes place not directly above the metal atom in the ledge or kink site but rather above the field adsorbed gas atom. So far it is not quite clear whether the field ionization above adsorbed gas atoms introduces errors if one determines atomic positions from FIM micrographs although most experiments indicate that FIM images may be interpreted as if there were no field adsorbed gas atoms present.

Considered from the practical point of view, field adsorption, especially of hydrogen, is of great benefit for imaging the non-refractory metals in the FIM. A small addition of hydrogen to helium as image gas reduces the best image field by 50% compared to pure helium[28] and by 36% as compared to pure neon. This hydrogen-promotion effect also lowers considerably the field necessary to remove atoms from the tip surface by field evaporation (Chap. 2.4). The reduced evaporation field exposes the tip to a much lower field induced mechanical stress (Chap. 2.5) and, therefore, the probability of rupturing the tip is markedly decreased. In order to smooth the surface of non-refractory tips after their preparation, the tips are often field evaporated in a He-H_2 mixture until they have reached their ideal *field-evaporated end-form;* then the hydrogen-helium gas mixture is removed and further imaging is performed with neon since the presence of hydrogen often introduces artificial vacancies into the tip surface or/and spreads the distribution of mass to charge ratios (m/n) in quantitative atom probe experiments by forming metal hydrides. This leads to a reduced mass resolution of the atom probe which can not be tolerated in most investigations (Chap. 3.3). Although not yet completely understood, it is generally assumed that hydrogen-promoted field ionization results from a field adsorption of hydrogen above the ledge or kink atoms in the surface leading to a better thermal accommodation of the hopping He atoms[29]. The same explanation has been given and experimentally proved for the case of neon promotion in a neon-helium gas mixture[29], where neon is field adsorbed. However, neon promotion only results in a better contrast and enhanced brightness but does not reduce considerably the best image field or the field evaporation field. Neon-helium gas mixtures, therefore, do not play an important role in imaging non-refractory metals and alloys.

2.4 Some Aspects of Field Evaporation

Field evaporation provides the means to remove atoms in the form of positive ions from the surface of a FIM tip if a sufficiently high positive voltage is applied to the specimen. Since the onset of field evaporation occurs at a certain field strength, regions of higher field strength, i.e. regions having smaller radii of curvature, field evaporate earlier than those of lower field strength. This effect is used to smooth a freshly prepared FIM specimen on an atomic scale until all sharp surface protrusions or oxide layers retained from the specimen preparation are removed. The final field evaporated end-form of the tip usually yields – at least in the case of pure metals – a perfectly developed field-ion image. In order to obtain a stable field-ion image, it is of course necessary that the best image field, F_{BIF}, is lower than the field evaporation field, F_{FE}, i.e. $F_{BIF} < F_{FE}$. (From the experimental point of view it is more adequate to speak in terms of best image voltage, F_{BIV}, and evaporation voltage, F_{FEV}, rather than in terms of field.) If this requirement is satisfied for the material under investigation, then the necessary voltage increase to field evaporate surface atoms can be applied either steadily to the tip (*DC evaporation*) or in a more controlled fashion by applying a voltage pulse of a few nanoseconds duration with an amplitude identical to the difference between F_{FEV} and F_{BIV} (*pulse evaporation*). The latter mode is crucial to ToF and imaging – atom probe analyses.

Although field evaporation is so important to analytical field ion microscopy and is routinely used in all experiments the mechanism of field evaporation is by no means well understood. To explain the phenomenon of field evaporation and in order to calculate the onset field of field evaporation and the charge state of the field evaporated metal ions several models have been proposed in the past[30–32]. These models treat field evaporation as a *one-dimensional* desorption mechanism and differ essentially in the assumption whether the transition to the ionic state occurs before field evaporation from the surface (Schottky-hump model by Müller[33]) or afterwards (charge-exchange model[31]). The most widely accepted charge-exchange model, which was originally developed by Gomer and Swanson[31] to describe field desorption of covalently bound adatoms from metals, has been adopted by Tsong[33] for field evaporation of metal atoms. In this model, field evaporation of an atom A from a metal surface M is considered as a charge rearrangement process between the systems $A + M$ and $A^{n+} + M^{n-}$. This transition to a n-fold positively charged ion A^{n+} occurs at a distance x_c from the metal surface, where the potential curves $U_a(x, F)$ and $U_i(x, F)$ for the $A + M$ and $A^{n+} + M^{n-}$ system, respectively, intersect in the presence of an applied field F (Fig. 2.7). With reference to Fig. 2.7 the activation energy $Q_n(F)$ is then

$$Q_n(F) = U_i(x_c, F) - U_a(x \approx 0, F) \, , \tag{2.2}$$

where $U_i(x_c, F)$ denotes the ionic potential energy at x_c, and $U_a(x \approx 0, F)$ the atomic potential energy; for the sake of simplicity the equilibrium position of the surface atom in

12

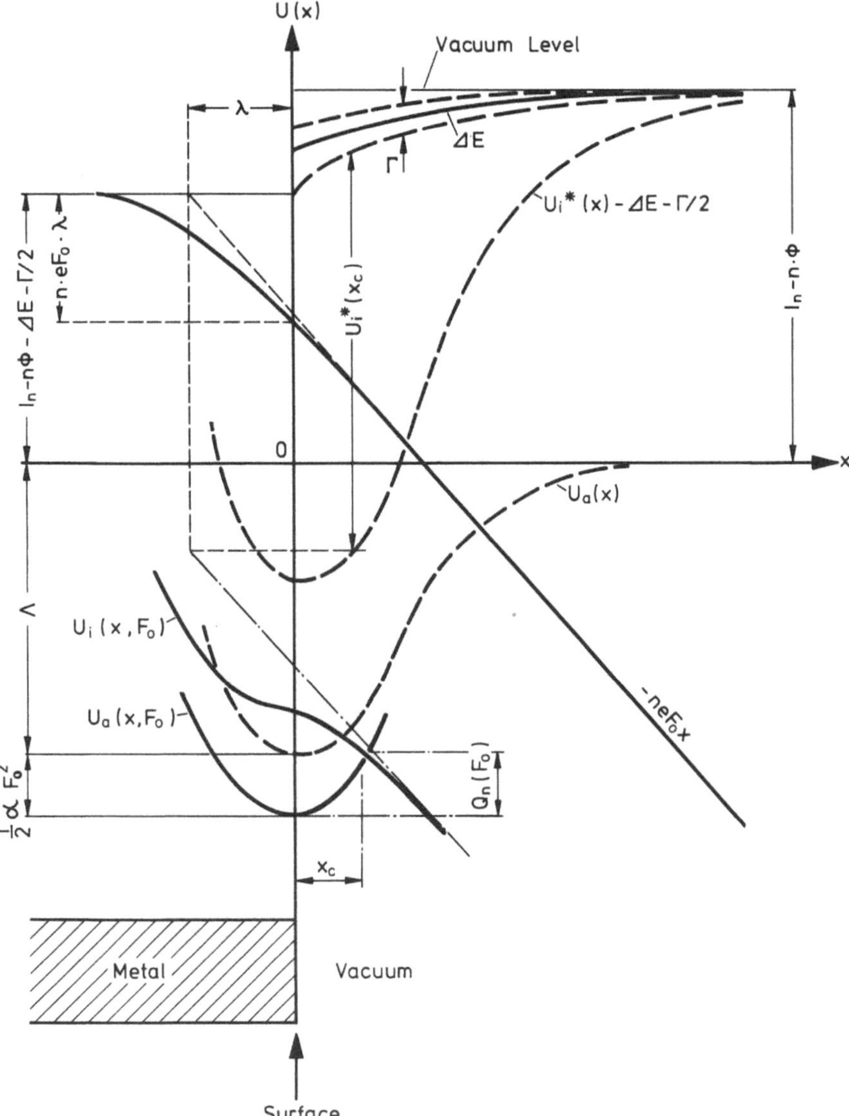

Fig. 2.7. Schematic potential-energy diagram for the process of field evaporation according to the charge-exchange model. The *continuous curves* show the potential energy for the neutral state $A + M$ ($U_a(x)$) and for the ionic state $A^{n+} + M^{n-}$ ($U_i(x)$) in the presence of the applied field F_0. The *dashed curves* show the corresponding potential energies in zero field. (After Ref. 29 and 34)

the presence of F is assumed at $x \approx 0$, i.e. at the metal surface. Omitting terms for hyperpolarizabilities, the atomic potential can be expressed as

$$U_a(x \approx 0, F) = -\left(\Lambda(F) + \frac{1}{2}\alpha F^2\right)$$

and is essentially determined by the polarizability α of the surface atom and by the sublimation energy Λ, which is altered in the presence of F by comparison to the zero-field case, because of a shift of free electronic charge in the presence of high electric fields. Following Tsong[33], and the potential energy scheme shown in Fig. 2.7, the ionic potential can be written as

$$U_i(x_c, F) = I_n - n\phi - \left(\frac{\Gamma(x_c)}{2} + \Delta E(x_c)\right) - neF(x_c + \lambda) + U_i^*(x_c).$$

Far away from the metal surface the potential energy of the n-fold charged ion is $I_n - n\phi$ with respect to the atomic state; I_n is the n-th ionization potential and $n\phi$ the energy gained from the transfer of n electrons into the metal of work function ϕ. Γ and ΔF account for the broadening of the electronic level and the electronic level shift, respectively, in the atom undergoing field evaporation; both result from the interaction of the atom with the metal surface[31] and lower the ionic potential in the vicinity of the surface. In a positive applied field the energy of the ion is further reduced by $neF(x_c + \lambda)$, where λ is the penetration depth of the electric field into the metal surface; $U_i^*(x_c)$ accounts for the interaction potential of an ion with a metal surface and includes both the attractive image potential and the repulsive ion-core interaction potential.

Thus, the activation energy can be written as

$$Q_n(F) = \left\{I_n - n\phi - \left(\frac{\Gamma(x_c)}{2} + \Delta E(x_c)\right) - neF(x_c + \lambda) + U_i^*(x_c)\right\}$$
$$\left\{-\Lambda(F) - \frac{1}{2}\alpha F^2\right\}.$$

(2.3)

Due to variations in the local electric field across the surface of the field-ion tip, and also because of the dependence of Λ upon the co-ordination number of the atom to be field evaporated, i.e. upon its environment, Q_n also varies across the tip surface. For pure metals it is immediately evident that atoms forming sharp protrusions (Chap. 2.2) and atoms on kink or ledge sites (Fig. 2.4) have the lowest Q_n, and, thus, will be field evaporated first; in alloys the situation is sometimes more complex[9] resulting often in the preferential field evaporation of either solute or solvent atoms at a given field (Chap. 2.6.1). Furthermore, due to the probable orientation dependence of both Φ and Λ[9, 10], Q_n varies with crystallographic region resulting in a *field-evaporated specimen endform* which is not ideally hemispherical but rather is composed of segments with locally different radii of curvature (see also Chap. 2.7.1).

Recently Ernst[34] measured the activation energies Q_1 and Q_2 for field evaporation of Rh^{1+}- and Rh^{2+}-ions from a pure rhenium tip. In this work it was verified that field evaporation of singly charged Rh^{1+}-ions can be reasonably well described in terms of the charge-exchange model. However, based on the observation that $Q_1 = Q_2$, Ernst concluded that the doubly charged ions originally evaporate as singly charged ions, and are subsequently post-ionized to Rh^{2+}. These results, which have to be corroborated for other metals, imply that the energy diagram shown in Fig. 2.7 applies only to field evaporation of singly charged ions, and that higher ionization states are obtained by post-ionization of singly charged ions rather than by an n-fold ionization process at the intersection of $U_i(x_c, F)$ and $U_a(x \approx 0, F)$.

14

Field evaporation occurs either by thermal activation[30] at higher temperatures or by atomic tunneling[31] at very low temperatures. In the intermediate temperature range between 20 K and ~ 90 K at which most metals and alloys are usually imaged and field evaporated, the evaporation rate $k_e^{(n)}$ is determined by both thermal activation and tunneling. The rate has been shown for this case to be[31, 32]

$$k_e^{(n)} \approx \nu_{eff} \cdot \exp\left\{ -\frac{Q_n}{kT} + \frac{C}{(kT)^3} \right\} . \qquad (2.4)$$

The term $\propto \exp\{C/(kT)^3\}$, which accounts for the atomic tunneling, has been derived[9] by approximating the potential barrier at x_c (Fig. 2.7) by a triangular potential wall of height Q_n; this leads to the T^{-3} dependence in the exponent of the tunneling term. The constant C depends on the atomic mass of the field evaporating atom, as well as on the shape of the triangle. During field-ion microscopic investigations it becomes evident that the evaporation rate increases considerably with increasing temperature; i.e. in order to maintain a stable field-ion image at increasing T the imaging voltage has to be reduced. So far, systematic rate measurements have only been carried out at relatively high temperatures ($\gtrsim 80$ K)[32], where thermal activation is believed to be the rate-controlling mechanism; at low temperatures there are no systematic studies reported which have tried to verify the temperature dependence of the evaporation rate as predicted by Eq. (2.4).

It is common to all field evaporation theories that they deal with field evaporation from a perfectly clean surface. We have seen, however, that under the conventional conditions for operating a field ion microscope, at least in the high-field regions, where field evaporation first starts, image gas atoms and/or impurity gas atoms from the background of the vacuum system are field adsorbed. The presence of these gas atoms decreases the evaporation field by a few percent in the case of helium and neon, and by up to fifty percent in the case of hydrogen. Furthermore, Brenner and McKinney[35] demonstrated that the relative abundance of different charge states of field evaporated metal ions A^{n+} depends on the evaporation rate and on the gaseous environment during evaporation. These effects and the formation of metal/inert-gas complex ions or metal-hydride ions as frequently observed in atom-probe spectra[35] are not adequately taken into account in all evaporation theories and have been used to explain the discrepancy between the experimentally and the theoretically determined field evaporation parameters[36].

Recently, Waugh et al.[37] investigated systematically the influence of various evaporation parameters such as tip temperature, gaseous environment and evaporation rate on the field evaporation characteristics of a wide variety of metals, including Fe, Al, Ni and Au, by means of the field desorption microscope and the imaging atom-probe. In these instruments the field evaporated metal ions are used themselves to form a desorption image as will be described in Chap. 3.5. Unlike conventional field ion microscopy no image gas is needed to form a field desorption image. (Sometimes field evaporation is synonymously termed field desorption although – strictly speaking – field desorption defines the removal of extrinsic surface species, e.g. field adsorbed image gas atoms, chemisorbed molecules or vapour deposited metal atoms. In the context of this article, field desorption is always meant in the sense of field evaporation of lattice atoms.) Therefore, field evaporation can be studied in the absence of field adsorbed image gas atoms with the desorption microscope or the imaging atom-probe. The desorption images of Waugh et al. were obtained by *continuous* field evaporation of some tens of successive lattice planes. Since the field-evaporated ions which form the integrated desorption image originate from different monolayers and since the number of atoms removed from unit surface area is uniform it was to be expected that an integrated desorption image would be essentially of uniform brightness and should not show any crystallographic features[37]. However, desorption images integrated over ten to a few hundred field evaporated monolayers of various metals all exhibit a fine structure consisting of bright (e.g. W, Rh, Ir, Au, Ni, Re) or dark zone lines (e.g. Al, see Fig. 2.8) with superimposed intricate networks of faint dark lines; the centres of low-index poles often appear dark and are frequently surrounded by dark concentric rings (Fig. 2.8). The striking appearance of the fine structure in the integrated desorption pattern has been explained by Waugh et al.[37] on the basis of a phenomenological field evaporation model which includes short-range migration of the atoms in specific crystallographic directions prior to their evaporation. According to this model an atom originally on a kink site position on a low-index plane will roll up over the edge of its lattice plane to a site where it is exposed to a higher field (Fig. 2.9); there it will be field evaporated. This movement will be possible if the activation energy Q^M for migration over the plane edge is sufficiently low to be overcome either by thermal activation or by atomic tunneling. The magnitude of Q^M is a function of the lattice binding energy which can be maintained during migration and thus depends on the number of nearest neighbour bonds involved.

Based on this hypothesis, Waugh et al. concluded that only the kink site atoms, which are shaded in the ball model of a {200} fcc lattice plane show in Fig. 2.10, can move over the plane edge along a $\langle 100 \rangle$ direction (indicated by arrows), since during migration along this path they remain in a nearest-neighbour relationship with two atoms in the top

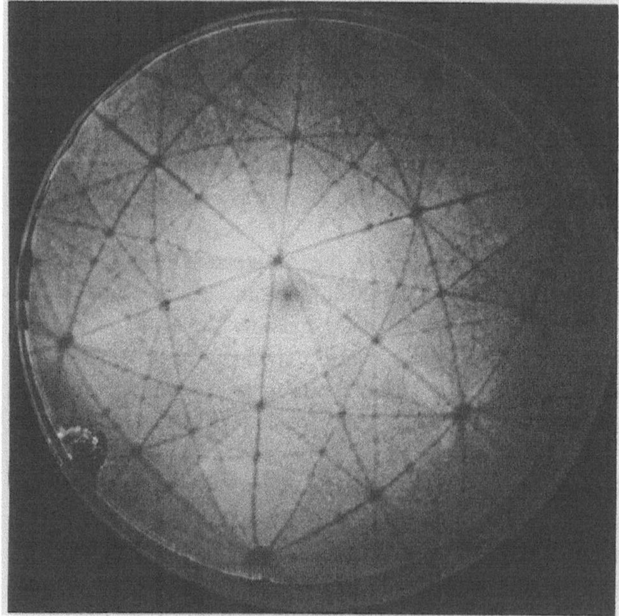

Fig. 2.8. Non-gated multilayer desorption image of aluminium; tip temperature 10 K. The pattern of *dark lines* is quite apparent. The pole close to the center is (111). (Courtesy A. R. Waugh)

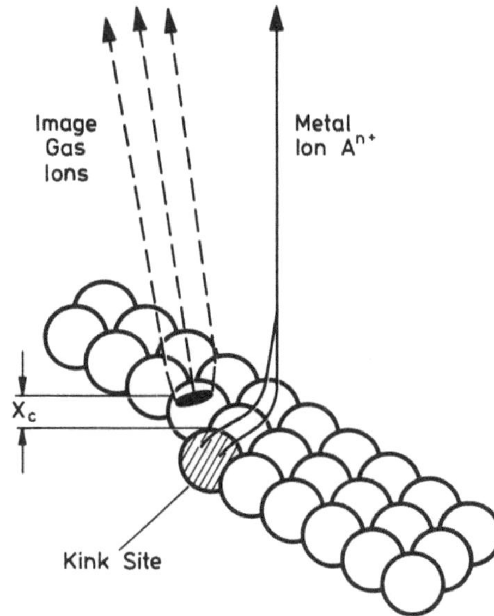

Fig. 2.9. Illustration of the rolling motion of the kink-site atom up over the edge of the ledge where it will finally field evaporate. Also shown are the ionization disc above the evaporating atom and the different trajectories for the image gas ions and the evaporated metal ion. (After A. R. Waugh et al.[26])

Fig. 2.10. Plan view of a quarter segment of the {200} plane of a fcc metal. According to the model of Waugh et al. [37] only the shaded kink-site atoms can move over the plane edge (see Fig. 2.9) along the arrowed directions prior to field evaporation. The zone-decorating atoms (A) are not able to move because of too high an activation energy. Atoms labelled B and C move towards the ⟨110⟩ zone lines (Z.L.) from two different directions

plane and one in the plane beneath. By contrast, such movement of a kink site atom at the end of a close-packed atomic row (labelled A in Fig. 2.10) involves contact with only one nearest neighbour, and is therefore associated with too high an activation energy for it to occur. (Examination of the evaporation characteristics of {200} regions of fcc metals confirms that these zone-decorating atoms (A) are, in fact, the most resistant to evaporation.) Due to this motion of the atoms along specific crystallographic channels prior to evaporation, the trajectories of ions evaporated from adjacent specimen areas along certain zone lines will overlap (e.g. atoms B and C in Fig. 2.10); the corresponding zone lines ([011] and [01$\bar{1}$] in Fig. 2.10) will then appear twice as bright in the multilayer desorption pattern in comparison to regions where the ion yield is not doubled. Dark zone lines will result when the atoms move away from the zone line prior to evaporation.

In cases, where the local surface topography does not allow an energetically favourable path (e.g. for atoms labelled A in Fig. 2.10), the atoms will evaporate at a higher field than their neighbours, often in a higher ionization state, as has been experimentally observed.

In the case of thermal activation, the energy barrier Q^M to migration is given by

$$Q^M \approx \Lambda_1 - \Lambda_2 + \frac{1}{2}\,\alpha_1 F_1^2 - \frac{1}{2}\,\alpha_2 F_2^2\,. \tag{2.5}$$

Here $\Lambda_1 - \Lambda_2$ describes the difference between the zero-field binding energy at the original lattice site (1) and at the saddle point between (1) and the high-field exposed ledge site (2); $1/2\,\alpha_1 F_1^2$ and $1/2\,\alpha_2 F_2^2$ are the corresponding polarization energies at the two sites. It is important to mention that the distances covered by the migrating atoms prior to field evaporation are less than one lattice constant.

The field evaporation model described above is a three-dimensional model since it involves migration. The proposed model can explain *qualitatively* many of the fine structures observed in the desorption patterns and is further supported by previous observations of the diffusion of surface atoms prior to field evaporation[35, 39, 40]. However, the reasons for the observed differences in the details of desorption patterns from different metals of the same crystal structure, e.g. fcc Au, Ni and Al, are not well understood, but have been assigned to variations in the interatomic forces of the different metals. Furthermore, the influence of the presence of image gas upon the field evaporation characteristics, e.g. alteration of the path of migration or changes in the crystallographic distribution of differently charged ions, remains unclear.

Due to both the short-range migration of the field evaporating ion and a parallax effect[18] resulting from a different location of the imaging-gas ionization zone (at x_c in Fig. 2.9) and the field-evaporation site, the trajectories of the field evaporated metal ions do not exactly coincide with those of the corresponding imaging gas ions (Fig. 2.9); therefore, an aiming error is encountered if an individual atom, which is preselected in the FIM image, is aimed with the atom probe for its chemical identification (Chap. 3.1). The existence of this aiming error was first noticed by Brenner and McKinney[18, 35] who reported that the detection of field evaporated ions consistently occurred *before* the evaporating net plane edge was imaged over the probe hole of the atom probe.

Fortunately, in practical applications of field-ion microscopy, field evaporation can be handled in many cases more or less empirically in the sense that one usually determines the onset of field evaporation by slowly increasing the voltage above the best image voltage F_{BIV} until the first atoms can be seen to evaporate. Since $F_{BIF} < F_{FE}$ is required, it is often useful to know approximately the evaporation field F_{FE}, at least for pure metals, in order to determine the specific image gas which satisfies $F_{BIF} < F_{FE}$ or in order to anticipate preferential field evaporation of particular species in an alloy (Chap. 3.3). In Table 2.1 some values of the calculated evaporation fields for various elements are listed and compared with the observed onset-field for evaporation. The calculations are based on Eq. (2.3) setting $Q_n = 0$ and neglecting the polarisation term as well as ΔE and Γ[9]. In some cases the agreement between calculated and observed values is rather good (e.g. for Co^{2+}, Ni^{2+}, Nb^{2+}, Pt^{2+}). However, for many elements the theory predicts different values (e.g. for Al) and even different charge states for the evaporated ions than observed experimentally. This becomes especially evident in the case of Cu, where Cu^{1+}

Table 2.1. Calculated field-evaporation fields F_{FE} at 0 K for 1-fold and 2-fold charged ions (the most abundant ions are underlined); best image fields F_{BIF} for some common image gases and the mass-to-charge ratios m/n for the most abundant isotopes of the listed elements (all fields in V/nm)

Element	Calculated[a] F_{FE}^{1+}	F_{FE}^{2+}	Observed[b]	Observed	Ref.	F_{BIF}[a]	m/n (amu)
Be	54	38.4	34	Be^{1+}, Be^{2+}	29	He 44	9, 4.5
C	142[c]	103[c]	–	C^{2+}, C^{+}, C_3^{2+}, C_2^{+}, C_3^{-}	42	Ne 34.5	6, 12, 18, 24, 36
Al	16.1	32.8	33	$\underline{Al^{+}}$, Al^{2+}	41, 43	Ar 19	27, 13.5
Si	44.3	31.7	30	$\underline{Si}^{2+\,d}$	44	H_2 18.8	14
Ti	35.3	23.3	25	$\underline{Ti}^{2+}$, $Ti^{3+\,d,\,e}$	45		24, 16
V	38.4	25	–	$\underline{V}^{2+}$, V^{3+}	44		25.5, 17
Cr	26.8	26.8	–	Cr^{1+}, $\underline{Cr}^{2+}$	29		52, 26
Mn	28[c]	28[c]		Mn^{2+}	42		27.5
Fe	40.6	31.8	36	Fe^{1+}, $\underline{Fe}^{2+}$, $Fe^{3+\,e}$	45		56, 28, 18.7
Co	41.8	34.9	37	Co^{2+}, Co^{3+}	33		28.5, 19.7
Ni	33.6	33.0	36	$Ni^{2+\,d}$	46		29
Cu	30.8	33.0	30	Cu^{+}, $Cu^{2+\,d,\,e}$	47		63, 31.5
Nb	52.6	34.8	40	$\underline{Nb}^{2+}$, Nb^{3+}	46		46.5, 31
Pt	50.4	44.2	47.5	Pt^{2+}	46		97.5
Ag	23.1	44.2	–	Ag^{+}	46		108
Au	40.2	48.5	35	Au^{+}, Au^{2+}, $Au^{3+\,e}$	48		197, 98.5, 65.7
Pd	36.3	40.8	–	Pd^{+}	45		~ 106
Mo	47.7	45.2	45	Mo^{2+}, Mo^{3+}, Mo^{4+}	7		49, 32.7, 24.5
B	64	79	–	B^{+}, $\underline{B}^{2+}$	45		11, 5.5

[a] Data calculated by Müller and Tsong and compiled in Ref. 9
[b] The observed evaporation fields have been compiled in Ref. 9, except for Al, whose value is given in Ref. 43
[c] From Ref. 10, calculated on the basis of the image force model
[d] Ions evaporated from alloys rather than pure metals
[e] Ratio between M^{1+} and M^{2+} increases with temperature and may depend on the vacuum conditions

is predicted by theory because of its lower evaporation field. In contrast, at lower temperatures (< 30 K) more than 50% Cu ions are doubly charged[47]. As can be seen from the fourth column of Table 2.1 all less refractory transition metal atoms evaporate around 35 V/nm, which is approximately the best image field for neon. The most abundant evaporated ions are doubly charged, no matter whether the ions result from pure metals or from alloys. Practically, it has turned out that all less refractory metals can be imaged with neon provided the tip temperature is lowered into the range of 80 to 20 K.

2.5 Mechanical Stresses Exerted by the Applied Field

During imaging and field evaporation, the field ion specimen experiences a mechanical stress $\sigma = F^2/8\pi$ exerted by the electrical field strength F[51, 52]. Since F varies across the tip surface, the associated mechanical stress also varies and, in order to determine the stress tensor at each point within the apex region of the specimen, the local electrical field distribution must be known. By assuming that the variations in regional brightness of the field ion image are mainly caused by variations in the local radius of curvature and, hence, in the local electrical field (Chap. 2.7.1), Eaton and Bayuzick[49] reconstructed the field distribution in the surface of a [011] oriented tungsten tip. By this means and by employing the finite-element method, they succeeded in determining the stress tensor at each point within the apex region of this tip; the computed shear stresses σ_{rz} are shown in Fig. 2.11. Beyond this, Eaton and Bayuzick have also shown[50] that not only the specimen shape but also the instrumental design parameters (e.g. whether or not an accelerating electrode located about 10^4 tip radii before the tip is used, or in which way the specimen is mounted onto the specimen holder and how it is connected to the cold finger, etc.) may significantly change the stress state in the specimen. Because of these additional complications, a precise calculation of the stress state is difficult, if not impossible.

From Fig. 2.11 it can be seen that the shear stress in the vicinity of the (111) pole is considerably higher than the yield stress σ_y for plastic flow. Fortunately, most field-ion specimens are perfect crystals, which are small enough to contain no dislocations. Therefore, in the absence of dislocations, the specimens do not fracture at positions where the field induced stress exceeds the yield stress σ_y. However, it is frequently observed in dilute Cu and Fe alloys that slip in the shank of the specimens occurs as the voltage is raised[53, 45]. This slip often leads to a complete fracture of the specimens; in some instances more than eight out of ten specimens fail[53]. It seems that Fe and Cu specimens often retain dislocations (probably introduced during specimen preparation) and, there-

Symbols	σ_{rz} (in $10^2 \cdot$ MPa)
A	4.5
B	3.6
C	2.7
D	1.8
E	0.09
F	0.0
G	-0.09
H	-1.8
I	-2.7
σ_{th}	~260
σ_y	~0.52

Fig. 2.11. Contours of computed shear stresses σ_{rz} within the apex region of a tungsten tip during imaging in helium at 14.5 kV. For comparison the ideal theoretical shear stress σ_{th} and the observed yield strength σ_y for pure tungsten are included (After H. C. Eaton and R. J. Bayuzick[49])

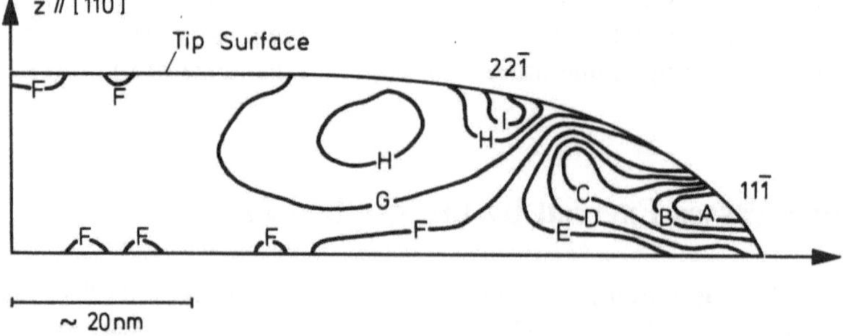

fore, are very susceptible to yielding. It should, however, be emphasized that, to our knowledge, there exists no transition metal which could not have been imaged in the FIM just because it always yielded before the best image field had been reached.

If yielding does not occur, the inhomogeneous field-induced stress will result in any case in a large negative pressure p associated with a non-uniform volume dilation ΔV in the apex regions; ΔV is of the order of 3% for most transition metals imaged with neon. This field induced pressure and its associated non-uniform volume expansion leads to some important consequences in the quantitative analysis of FIM experiments, particularly in studies of bulk diffusion performed during field-ion imaging, e.g. the migration of radiation-induced self-interstitial atoms (SIA) to the tip surface during the course of an *in-situ* recovery experiment[54, 55] (Chap. 4.3.4). In this case, the field induced negative pressure p affects the diffusivity of the SIA by changing the enthalpy change of migration ΔH^m according to

$$\Delta H^m = \Delta U^m + p\Delta V^m \tag{2.6a}$$

Here ΔU^m and ΔV^m denote the internal energy change of migration and the change in the volume of migration of a SIA, respectively. In practice[55, 56], the tip is usually considered as an elastically isotropic sphere and p is approximated by the hydrostatic pressure $F_{BIF}/8\pi$. Then 2.6a becomes

$$\Delta H^m = \Delta U^m - \frac{F_{BIF}}{8\pi} \cdot \Delta V^m \tag{2.6b}$$

Since ΔV^m is positive (for further discussion see Chap. 4.3.4), the effect of the electric field is to lower the enthalpy change of migration and, hence, to increase the diffusivity of a SIA.

The flux of migrating species through unit cross-section per time under constant pressure p and temperature T is proportional to the gradient in the concentration ∇c of these species. An additional pressure gradient ∇p, as present in an imaged field ion specimen, will lead to an additional flux which is proportional to ∇p[54]. All *in-situ* diffusion experiments in the FIM reported so far have dealt with the recovery of radiation-induced point defects. These defects have been produced (Chap. 4.3) within one radius of curvature beneath the tip surface. In this specimen region the pressure gradient ∇p is, however, very small[51, 52]. Therefore, the contribution to the flux resulting from ∇p can be neglected in these experiments compared to the contributions resulting from ∇c[54, 56].

2.6 Image Contrast of Alloys

2.6.1 Disordered Solid Solutions

Because of its atomic resolution, the optimum applicability of the FIM to the study of solid solutions is to investigate the distribution of solvent and solute atoms in binary alloys, to determine in short-range ordering or short-range clustering parameters. For this type of study it is necessary to identify and distinguish between the two atomic species. Unfortunately, this became only possible in a few dilute alloying systems, such as Pt-Au[57, 58], Pt-Ni[57, 58] and to some extent in various diluted binary Ir- and Pt-base alloys with additions of Zr, Rh, Pt, Ru[59] and W, Pd, Co, Ni, Au[60], respectively. In none of these binary alloys is species recognition straightforward. Furthermore, it is quite a common experience that the good image contrast and the atomic resolution in the FIM patterns, characteristic of pure metals, deteriorate significantly in most alloys with increasing solute content (Fig. 2.12 a). These irregular images no longer allow any detailed interpretation.

The reason for this image deterioration is not well understood. The most important process in alloy image formation seems to be the *"selective evaporation"* mechanism[59, 61, 62] of the solvent and the solute atoms. In the case of "selective solute evaporation" the FIM patterns should reveal a "dark site contrast" indicating a vacant site at the position of the selectively evaporated solute atom. Indeed, the dark site contrast has been found in several binary alloys, e.g. in Pt-Au (0.62 and 4.0 at% Au[57, 58]) and in Pt-Ni (0.15 and 0.65 at% Ni[57, 58]). However, Chen and Balluffi[58] have shown that not each dark site is necessarily associated wih the exact position of a Au or Ni atom. Besides the dark spots which represent the atomic sites of Au or Ni (termed α-dark spots), they observed additionally several dark spots (termed β-dark spots) which were associated with preferentially field evaporated solvent Pt atoms lying in the proximity of solute Au or Ni atoms. This observation clearly demonstrates that selective evaporation is not an inherent feature of one particular atomic species in the alloy but also depends on the local environment of the atoms bound in the surface of the tip. Beyond this, it has been established[63] that even in well annealed *pure* Pt specimens some crystallographic planes (e.g. {102}) yield a black site concentration of up to 2 at% due to a local *preferential field evaporation* of Pt atoms. Nevertheless being aware of the contrast problems and avoiding any counting of dark sites on {102}-planes, Chen and Balluffi succeeded in differentiating between the α-dark spots, i.e. the real solute sites, and the β-dark spots by carefully watching their appearance during the course of controlled pulsed field evaporation. The concentration of α-dark spots agreed within 5% with the nominal solute concentration in all above mentioned Pt based alloys (see Chap. 4.2.4).

A similar type of study in various Pt base alloys by Dubroff and Machlin[60] yielded a considerable higher concentration of dark spots than corresponded to the particular nominal solute concentration. This is most likely due to the fact that each vacant site appearing in the FIM pattern was attributed to a solute atom. This investigation clearly stresses the point that there exists no one-to-one relation between the apparent vacant site and a known atomic species in the alloy.

The solute additions Rh, Rt, Zr, Ru, Pt and W in binary Ir based alloys[61] and Mo in Fe[64] exhibit a "bright spot contrast", which can result either from *"preferential retention"* or *"selective ionization"* of the solute atoms. In the case of preferential retention the solvent atoms evaporate at a lower field than the solute atoms. Thus, during the course of field evaporation, the solute species can be retained in sites having lower than kink site co-ordinations, e.g. in isolated sites (see E and R in 2.4 a), where they give rise to bright image spots mainly because of their geometrical field enhancement. If only preferential retention is responsible for the bright spot contrast, the solute atoms can not be differentiated in contrast from matrix atoms as long as they are still close to the centre of a plane, i.e. bound in higher co-ordination sites.

It has been predicted[61] that some solute species undergoing preferential retention are already imaged in a bright contrast by "selective ionization" whilst they are still bound in high co-ordination sites. However, selective ionization, which is considered to be essentially due to a smaller charge density at the solute site and, hence, an enhanced electric field above that site, has been observed only in few alloys (e.g. Ir-W[59]) and it seems that geometrical effects, i.e. the extent by which an atom protrudes from the surface of the tip, rather than electronic effects dominate in imaging a particular alloy constituent.

Again, as in the case of alloys showing black-spot contrast, it is impossible to obtain the solute concentration of the particular alloy by just counting the number of bright spots since the selective retention mechanism which is responsible for the alloy contrast is strictly dependent on the local electrical field. Solute retention and, therefore, bright spot contrast is most apparent in the low-field regions of the surface while very little is seen in the high-field areas. Additionally, bright spots are often associated with interstitial impurities such as oxygen and nitrogen making the determination of the substitutional solute concentration from the FIM pattern, especially in bcc-alloys, even more complicated if not impossible.

There have been some attempts to predict the contrast features which are to be expected in various binary solid solutions[61, 65], e.g. to predict the occurrence of either preferential evaporation or retention or preferential ionization, but none of these approaches can explain all previously observed alloy contrasts. In practice, such a prediction, even if correct, would be only of academic interest since it is experimentally not possible to discriminate *a priori* between the contrast of a real solute atom and that of an artifact. Therefore, to investigate reliably short-range ordering or clustering by means of FIM, a very detailed knowledge of the contrast features and the artifact level of the particular alloy has first to be obtained from various control experiments. These control experiments sometimes can be extremely tedious. This fact might explain why the work of Chen and Balluffi[58] (see also Chap. 4.2.4) remains up to now the only reliable FIM-investigation dealing with short-range ordering and clustering phenomena.

2.6.2 Long-Range Ordered Alloys

Due to selective evaporation of either species, the field-ion images of most more concentrated solid solutions are structureless and the crystallographic regularity, inherent to the FIM patterns of pure metals, is no longer achieved as is shown in Fig. 2.12a for a disordered Ni-14 at% Mo alloy[66]. In contrast to the disordered solid solution, the same alloy in the ordered bc-tetragonal Ni_4Mo state gives a FIM image which is comparable in regularity to that obtained from a pure metal (Fig. 2.12b). The reason for this drastic improvement in image quality after ordering is the fact that the two atomic species are now regularly distributed on the superlattice and that only the Mo atoms are imaged; the Ni atoms, however, are not visible. Based on the measurement of the local radii of curvature and, hence, the field evaporated end-form of an ordered Ni_4Mo tip, Yamamoto et al.[67] rule out, that the invisibility of the Ni atoms is due to a preferential field evaporation but is rather due to a smaller ionization probability above the Ni atoms as compared to the Mo atoms. The same contrast features have been obtained for an ordered (orthorhombic) Ni_3Mo alloy[68]. In ordered Ni_2Cr alloys only the Cr atoms give rise to image spots on the field-ion micrograph as has been found by Taunt and cowork-

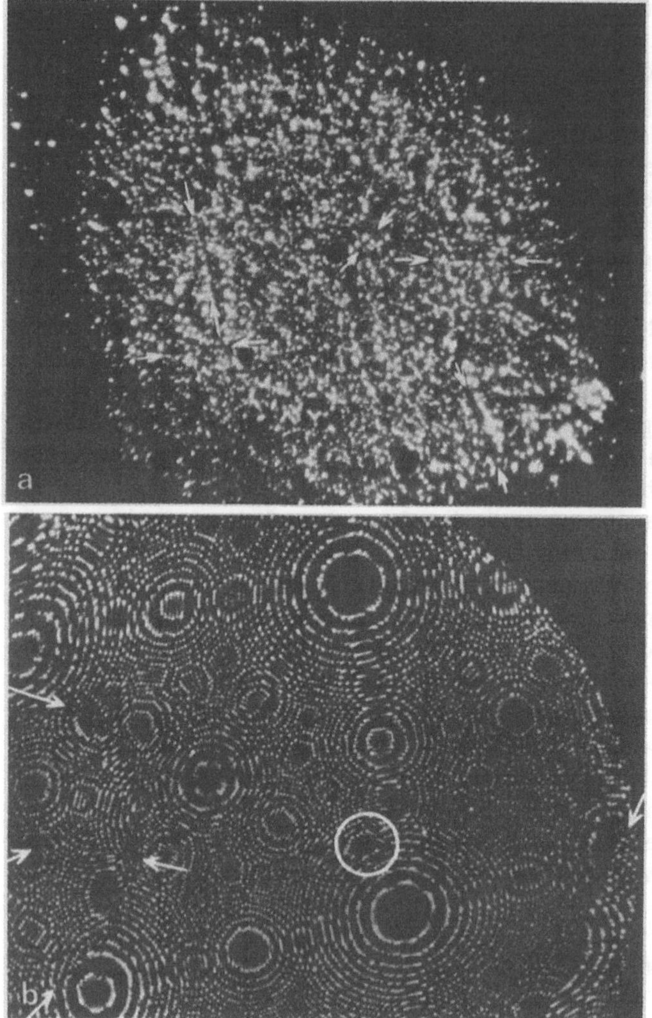

Fig. 2.12 a. Helium field-ion image of disordered Ni 14 at% Mo after quenching from 850 °C (21 K, 17.5 kV). **b)** Ordered Ni₄Mo after annealing at 750 °C (16 K, 30 kV). *Arrows* indicate domain boundaries (After R. W. Newman and J. J. Hren[66])

ers[69–71]. Therefore, in a superlattice plane stack, where two successive Ni planes alternate with a pure Cr plane, only every third plane yields an image ring. In a stack of fundamental planes, where each plane contains two Ni atoms to every Cr atom, each plane in the stack gives rise to an image ring, in which only the Cr atoms appear bright. In ordered Ni₂V having the same crystal structure (orthorhombic) as Ni₂Cr the contrast features are just the reverse since only the Ni atoms are imaged[70]; however, during the course of slow field evaporation of a stack of superlattice planes, the image rings of pure V planes become temporarily visible before they collapse.

Because of the drastic difference in the contrast features of a disordered and an ordered specimen, the FIM provides a useful mean to investigate the mechanisms of ordering phase transformations and consequently, several field-ion investigations on disorder-order reactions have been performed (see Chap. 4.4).

2.6.3 Decomposed Alloys

In most two-phase alloys the precipitates appear either (i) in a bright contrast or (ii) in a dark contrast with respect to the surrounding matrix. In both cases ((i) and (ii)) after some field-evaporation, a steady state end-form of the FIM tip evolves with the precipitates either protruding from the emitter surface (i) or being blunted (ii) as compared to the local radius of curvature of the adjacent matrix (Fig. 2.13). These local deviations in the radius of curvature above the precipitates are caused by a respecitively higher (i) or lower (ii) evaporation field than is necessary in order to field evaporate the surrounding matrix. Although not yet fully understood, it is assumed that the electric field variations associated with the topological variations in the surface due to the precipitate are responsible for an enhanced (i) or reduced (ii) ionization probability above the precipitates and, hence, a bright or dark contrast, respectively. Copper-rich precipitates in *Fe*-Cu show a dark contrast whereas gold-rich precipitates, which are more resistant against field evaporation than the surrounding matrix, appear bright in *Fe*-Au imaged with neon[72]. However, the contrast of a particular phase may depend also on the image gas and the temperature of the tip. For instance, the Cu-rich particles in an Fe-Cu alloy become bright if imaged with hydrogen instead of neon, and γ'-precipitates in a Ni-base superalloy exhibit bright contrast at a tip temperature of 80 K whereas hardly any contrast is to be seen at 30 K.

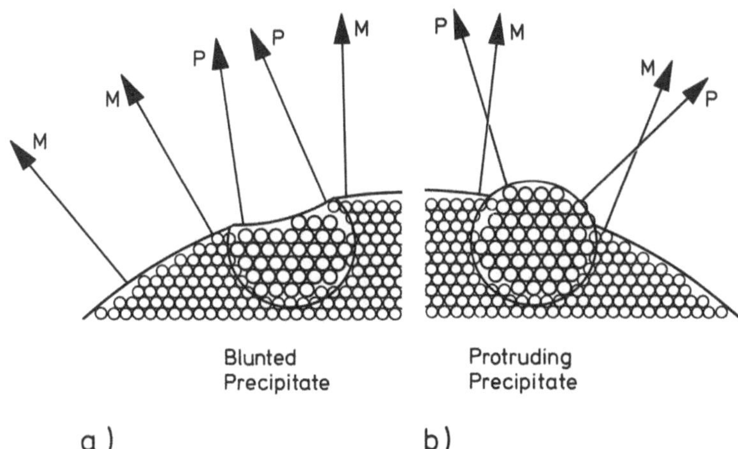

Fig. 2.13a, b. Surface topology around precipitates after the tip reached a steady state end-form. a) Blunted pricipitates appearing in dark contrast; b) protruding precipitates appearing in bright contrast. Schematically also shown are the trajectories of field-evaporating ions originating from either the matrix (*M*) or the precipitate (*P*)

The FIM images of some decomposed alloys, e.g. Cu-Ti, Co-Ta or Ni-Al, often do not reveal any contrast from precipitates and look similar to those of one-phase materials although atom-probe analyses and transmission electron microscopy clearly indicate the presence of particles of a second phase. Figure 2.14 a shows a rather regular field-ion image of a Cu-5 at% Ti specimen aged for 6 h at 350 °C; no discrete particles are to be seen. However, after part of the tip ruptured, all of a sudden the precipitated phase (dark contrast in Fig. 2.14 b) could be clearly distinguished from the bright matrix; after some field evaporation the contrast between the two phases disappeared again. A similar behaviour in the appearance of the phase contrasts has been reported for Co-Ta[73)] and Ni-Al. In aged Co-7.5 at% Ti no obvious signs of precipitation were visible with fairly sharp tips ($r_t \leq 40$ nm)[74)]. However, after blunting the tips to about 80 nm by continuous field evaporation, the Co_3Ti precipitates became finally visible.

These examples clearly demonstrate, that the absence of any precipitation contrast in the FIM pattern does not necessarily mean that the alloy has not yet decomposed into two distinct phases. In alloys where the precipitated phase is not visible one strictly has to

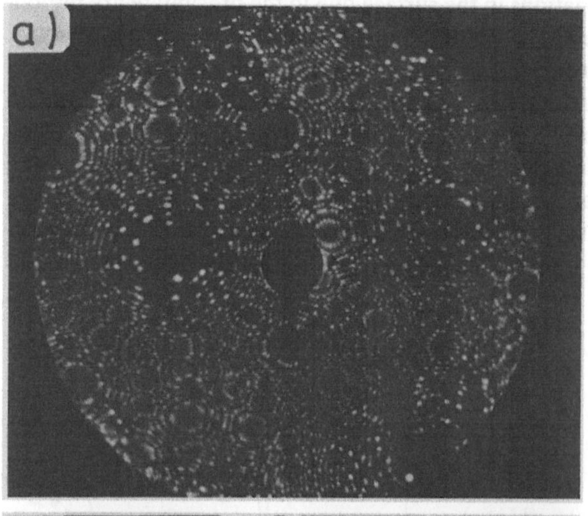

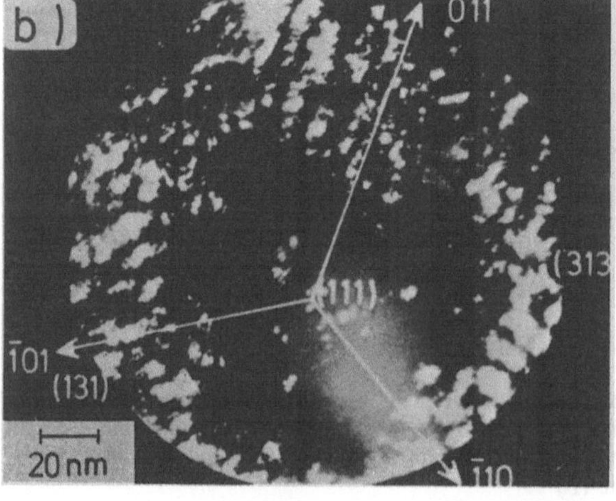

Fig. 2.14 a, b. Neon field-ion image od decomposed Cu-Ti specimens. a) Field-ion image showing no sign of phase separation; b) only after part of the tip ruptured the two phases can be clearly distinguished. (Aged for 6 h at 350 °C)

rely on atom-probe analyses (see Chap. 3.2) or on TEM, if possible, to investigate the state of phase separation.

The contrast features of field-ion images of decomposing alloys often also depend on the degree of decomposition. This is shown in Fig. 2.15a and 2.15b for a Cu-1 at% Fe alloy[75]. After aging the alloy for 100 min at 500 °C the matrix is still atomically resolved and the contrast from the precipitates is rather faint (a). After aging for 150 min, the atomic resolution of the matrix atoms (Cu) is lost but the contrast from the precipitates is drastically enhanced (b). The contrast enhancement in later stages of the precipitation reaction at the expense of image resolution of the matrix has been observed quite frequently in various two-phase alloys.

2.6.4 Vacancies and Interstitial Atoms

Because of its atomic resolution, the FIM has often been employed to investigate radiation-induced point defects such as single vacancies and self-interstitial atoms (SIA) (see Chap. 4.3). While a vacancy is simply imaged as a dark spot (one "only" has to ascertain that the black spot is not an artifact generated by preferential field evaporation; this can be achieved by watching the appearance of the black spot during controlled field evaporation) the contrast effect from a single SIA is more complex as was discussed in much detail by Seidman and coworkers for the case of an immobile $\langle 110 \rangle$ split SIA[76, 77]. Essentially they found three possible contrast effects which may be sequentially produced by one and the same SIA during the course of field evaporation of successive atomic planes. Due to its associated long range dilatational stress field, a SIA lying below the tip surface displaces the atoms in and near the tip surface. The SIA appears as a *"bright spot"* within a (111) plane of a bcc metal if one of the atoms *in the topmost* (111) plane is sufficiently displaced along [111]; this displacement increases the electric field locally and, hence, results in an enhanced image intensity of the displaced atom. If, however, an atom in the (111) plane *underneath the topmost* one is sufficiently displaced to increase locally the field above the topmost (111) plane, then the SIA manifests itself first by the appearance of a contrast which has been termed *"extra bright spot"* by Seidman[77], although its intensity is weaker than that of the "bright spot". The intensity of the "extra bright spot" is increased to about that of a "bright spot" if the atoms of the topmost (111) plane are removed by field evaporation. In some instances, the normal displacement of a surface atom in the (111) plane is sufficiently large to cause its preferential field evaporation. Then the "extra bright spot" or "bright spot", formerly associated with a SIA, will be replaced by a vacant site contrast, i.e. by a *dark spot*. In the latter case, it is again the carefully observed development of the contrast pattern during controlled field evaporation which makes it possible to distinguish between a dark spot contrast caused by the presence of a SIA or a real vacancy.

Amongst the contrast patterns of mobile SIAs, appearing on the surface of irradiated field ion tips during *in-situ* isochronal annealing experiments, the single bright spot contrast was reported to prevail by as much as 90%[77] (see also Chap. 4.3.4). In a few instances, however, also multiple-spot contrast patterns were observed, the double spot patterns of which were concluded to be produced by single SIA whereas pattern with more than two spots were assigned to SIA clusters.

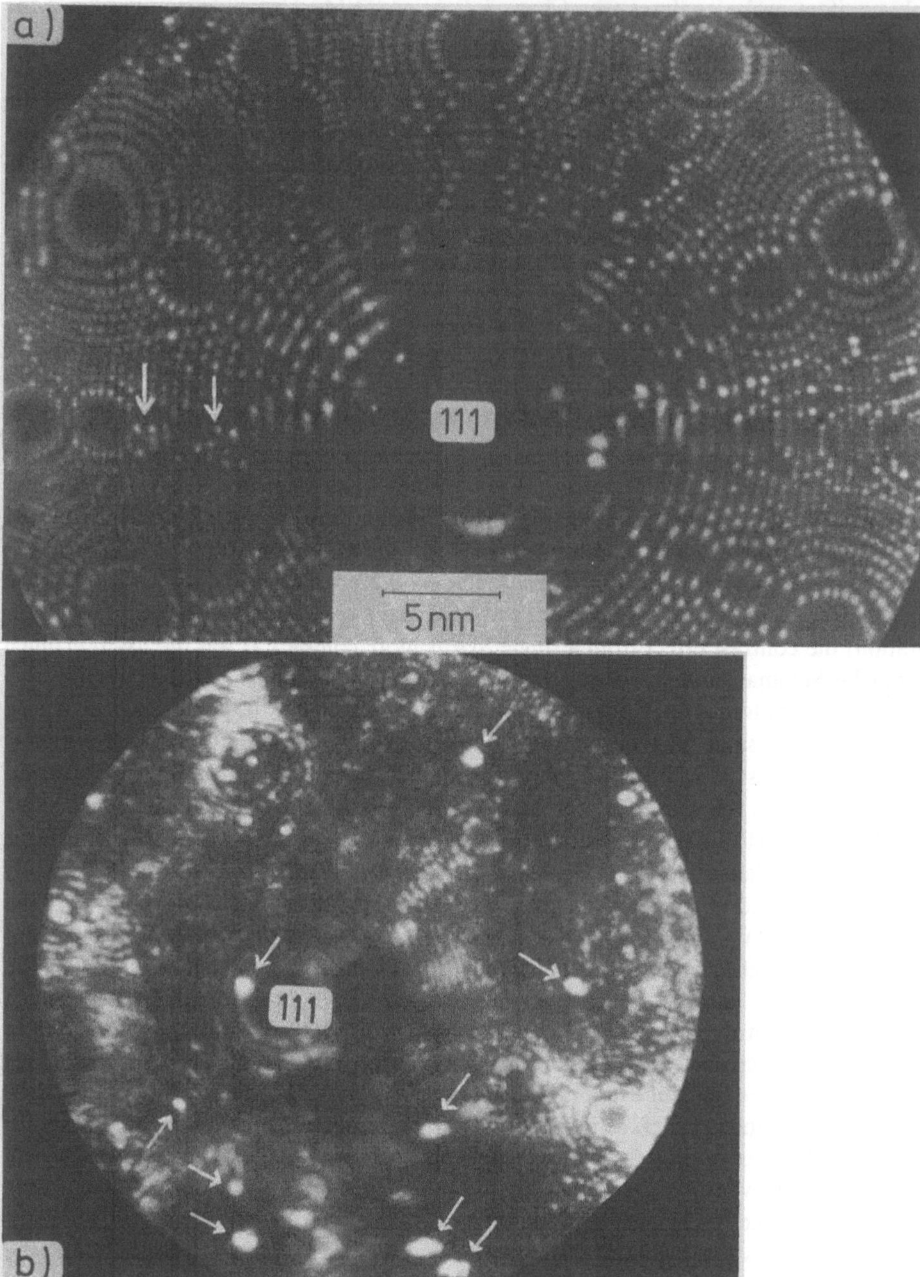

Fig. 2.15a, b. Neon field-ion image of Cu-1 at% Fe aged for 100 min at 500 °C; *arrows* indicate nuclei of a metastable iron-rich precipitating phase; **b)** the same alloy after aging for 150 min. Note the enhanced contrast of the Fe-rich precipitates as compared to **a)**[75].

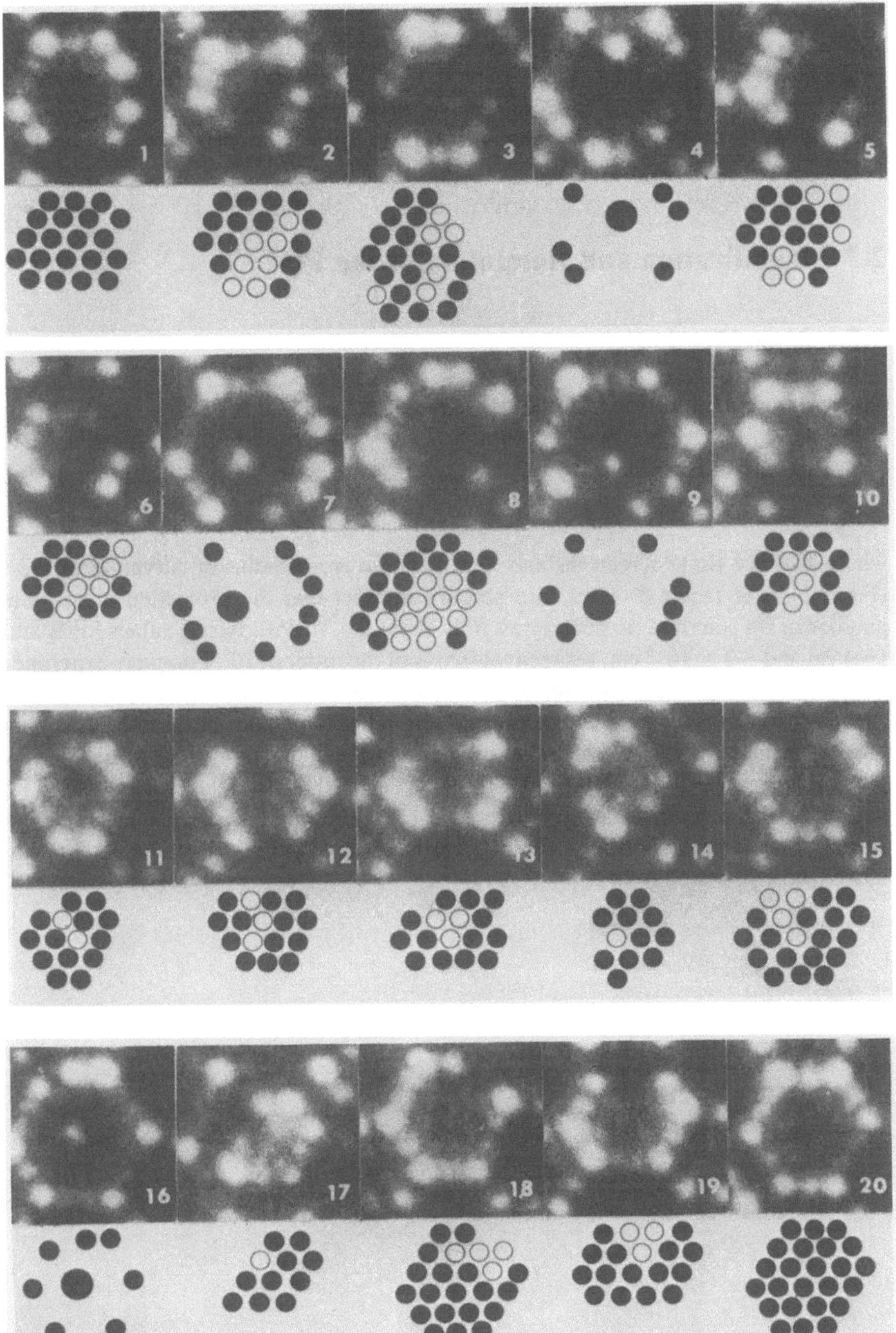

Fig. 2.16. Twenty FIM micrographs, out of a sequence of 4.4×10^3 recorded frames, showing a depleted zone in tungsten detected in the (111) plane at various stages of the atom-by-atom dissection process. The large *solid circles* in frames 4, 7, 9 and 17 indicate the SIA; the *open circles* indicate vacancies and the small solid *black circles* indicate lattice atoms[54]. (Courtesy D. N. Seidman)

Figure 2.16 shows the contrast patterns of some vacancies and SIAs which have been uncovered during the atom-by-atom dissection of a depleted zone in tungsten generated by bombardment with 20 keV W^+-ions at 18 K[54] (see also Chap. 4.3.2). During this dissection process 4.4×10^3 micrographs had to be recorded and analysed; Fig. 2.16 just shows 20 different micrographs taken at various stages of the dissection process.

2.7 Magnification and Resolution of the FIM

2.7.1 Magnification

The *average* magnification in the FIM is given by

$$\overline{M} = \frac{R}{\bar{r}_t} \cdot \beta' , \qquad (2.7)$$

where R is the tip-to-screen distance and $\bar{r}_t$ the average radius of curvature of the tip (Fig. 2.3). The factor β' takes into account the fact that the projection is not purely gnomonic; for standard tip geometries β' is about 0.6[78]. With typical values for R and $\bar{r}_t$ (~ 4 cm and $\sim 3 \times 10^{-6}$ cm, respectively) $\overline{M}$ is of the order of 10^6. For many experiments it is, however, necessary to determine the more accurate local magnification M. Then, according to (2.7), the local radius of curvature r_t has to be known. r_t can be determined directly from the FIM micrographs by counting the number n of rings between two previously identified poles (hkl) and (h'k'l')[79] (Fig. 2.17), (A method for indexing a FIM pattern has been described by Wilkes et al.[78]). By knowing the interplanar distance d_{hkl} as well as the angle θ between the two particular poles, the local r_t for that region is immediately obtained from

$$r_t = \frac{n \cdot d_{hkl}}{1 - \cos \theta} . \qquad (2.8)$$

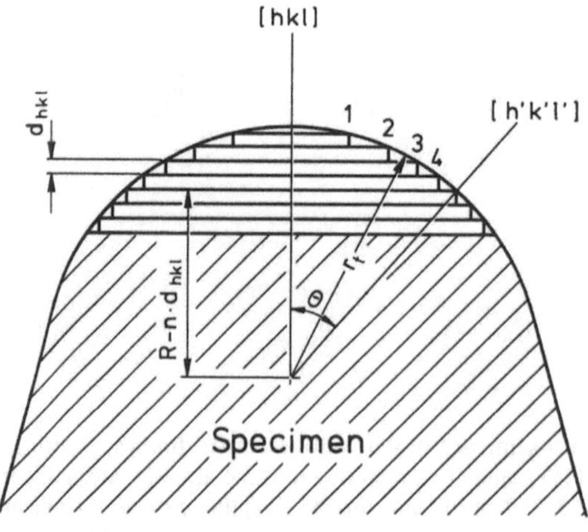

Fig. 2.17. Illustration of the determination of the local radius of curvature r_t (see text). (After M. Drechsler and P. Wulf[79])

This method works well as long as the FIM image shows a well developed ring structure, which is often not the case in alloys undergoing a precipitation reaction as we have seen in the previous chapters. With respect to the analysis of two-phase alloys in the FIM, the size distribution and morphology of the precipitates are the most interesting parameters. It is, however, in most practical cases not possible to determine the size of (visible) particles directly from a FIM micrograph just by using the average magnification or even by using the local matrix magnification of a region adjacent to the particular precipitate. Because of the localized (sometimes drastical) change in the radius of curvature (see Fig. 2.13 in Chap. 2.6.3) above precipitates there results a considerable unknown change in the magnification of the particular precipitate. If the precipitates are non-spherical, e.g. thin platelets or G.P. zones, the magnification may even be rather anisotropic[80].

Fortunately, the size of visible precipitates (and of other three-dimensional defects, such as clusters of point defects) can often be determined very accurately without knowing the magnification of the FIM micrograph by the method illustrated in Fig. 2.18 a. By controlled field evaporation several lattice planes of a previously identified pole (hkl) are successively removed. Whenever a precipitate or defect cluster becomes visible in the vicinity of this pole, the number n'_{hkl} of planes (hkl) is counted which have to be evaporated between the appearance and disappearance of the particle ("plane counting" or "persistence-size technique"). The dimension of the precipitate along the [hkl] direction is then immediately obtained as $d(hkl) \cdot n'_{hkl}$ with a depth resolution of ~ 0.2 nm. By recording the variations of the intersectional area of the precipitate with the tip surface during the dissection of the precipitate, the morphology of the three-dimensional defect can additionally be obtained quite accurately.

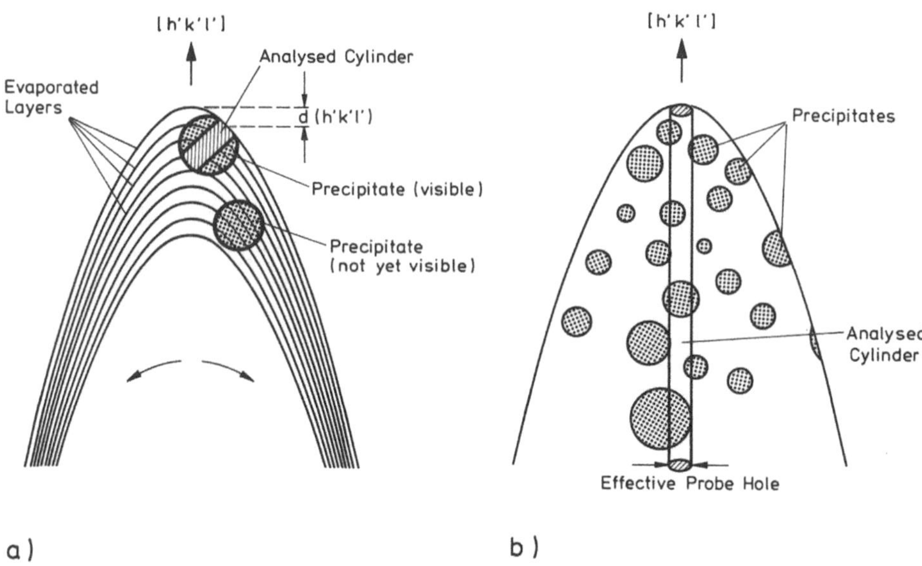

Fig. 2.18. a Lattice plane counting technique for the accurate determination of the size and morphology of visible precipitates (see text). **b)** The almost cylindrical volume analysed during an atom-probe experiment. Note that only some particles are completely cut by the cylinder

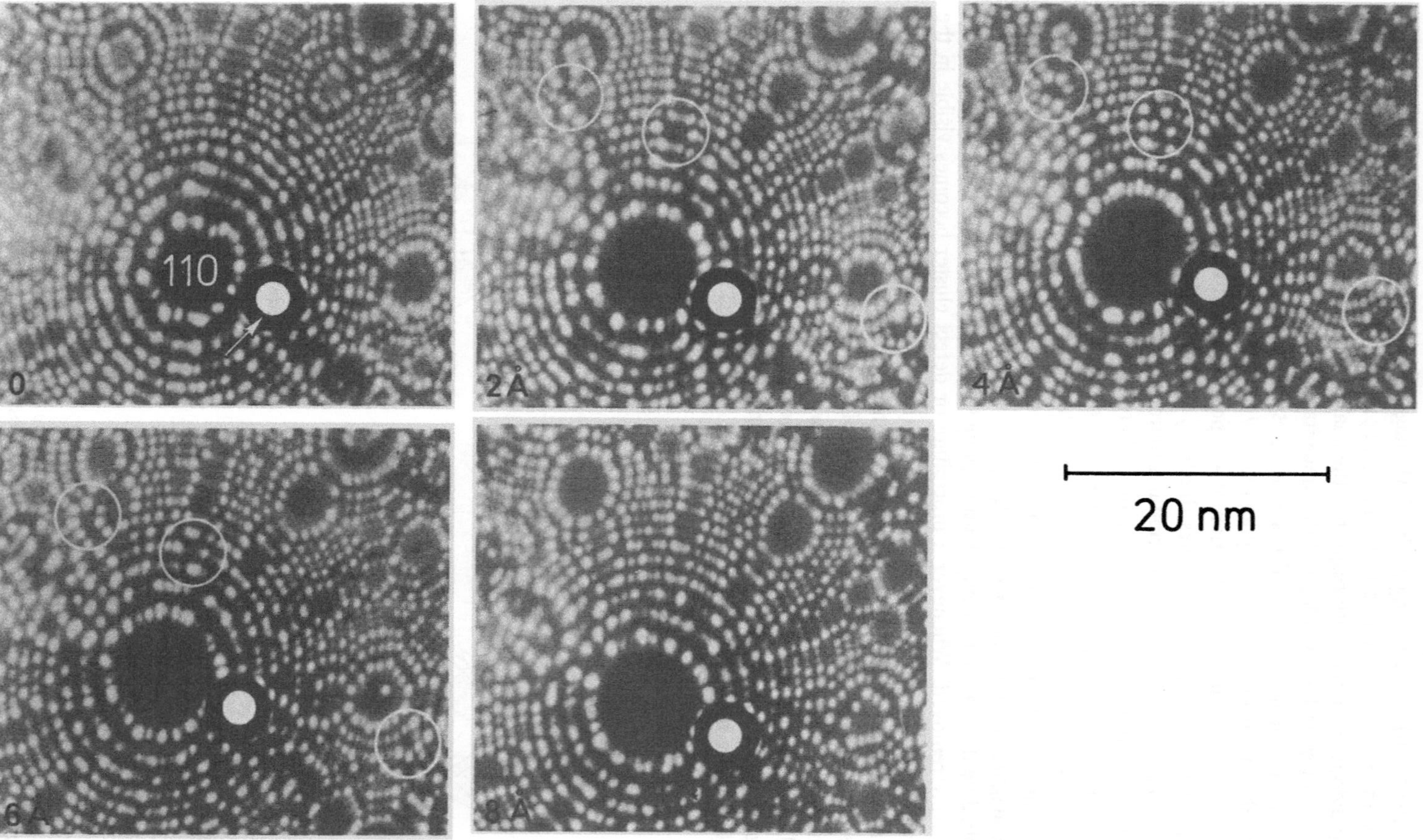

Fig. 2.19. Field-evaporation sequence through Fe-0.34 at % Cu irradiated to 3×10^{19} n/cm^2 (E > 1 MeV) at 288 °C. *Circles* indicate micro-voids. *Numbers below photographs* indicate cummulative depth of field evaporation. In the first frame the probe-hole is indicated by the *arrow*[81]

Figure 2.19 shows an example of a field evaporation sequence in neutron-irradiated Fe-0.34 at% Cu[81]. During the removal of four successive (011) planes (corresponding to 0.8 nm) three defects become visible. Two of them are persistent during the evaporation of three (011) planes while one persists for only two (011) planes.

2.7.2 Resolution

The resolution δ of the FIM is defined as the minimum size of the image spot of a single atom on the FIM screen at a given magnification M[9]. δ is essentially determined by three factors:

1) Ionization at x_c and beyond it (see Fig. 2.4b) occurs not only at a well defined point above the surface atom but rather takes place within an ionization disc of diameter δ_0; δ_0 which has been empirically determined to range from 0.15 to 0.35 nm[82], essentially limits the resolution at 0 K;
2) the broadening of the image spot caused by the uncertainty in the tangential momentum of the ion and the corresponding lateral velocity;
3) the broadening of the image spot size by the thermal lateral velocity of the imaging ion. Evaluation[9] and combination[82] of the three contributions yields the resolution δ of the FIM:

$$\delta = \left\{ \delta_0^2 + \frac{4\hbar}{\beta'} \left(\frac{r_t}{2\,meF_{BIF} \cdot \varkappa} \right)^{1/2} + \left(\frac{16\,r_t \cdot kT}{eF_{BIF} \cdot \varkappa \cdot \beta'^2} \right) \right\}^{1/2} \tag{2.9}$$

The factor $\varkappa$, ranging between 4 and 8, relates the applied voltage V_0 to the best image field F_{BIF} as $F_{BIF} = V_0/\varkappa \cdot r_t$. According to (2.9), the resolution is essentially determined by the tip temperature T and the tip radius r_t. Thus to obtain optimum resolution, cooling the tip to cryogenic temperatures is always necessary. From the previous sections it became apparent that (2.9) describes the optimum resolution which can be achieved only in either pure metals, some dilute alloys or long-range ordered alloys; for most two-phase materials, however, (2.9) becomes unimportant.

3 Single-Atom Mass Spectroscopy in Connection with FIM

3.1 The Atom-Probe FIM

Figure 3.1 shows the basic features and electronic circuitry of a time-of-flight atom-probe FIM, which we have designed for metallurgical applications. Essentially, the instrument consist of two parts: the FIM to image the specimen, and the time-of-flight (ToF) spectrometer (the so-called atom probe) in which single atoms, field evaporated from the FIM tip, are identified. The imaging-screen contains in its center a probe hole of about 2.5 mm in diameter, through which field evaporated ions, originating from the area covered by the projection of the probe-hole onto the specimen surface, i.e. the effective probe-hole area, enter the flight tube of the ToF spectrometer. At the end of the flight-tube the ions reach the single-ion detector.

The chemical identification of a single ion is achieved by first measuring its time of flight, t, and subsequently evaluating its mass-to-charge ratio m/n. For this purpose, the atoms are field evaporated at a well defined instant by superposing a high-voltage pulse of a few nanoseconds rise-time, to the tip which is held at the imaging voltage V_{DC}; a pulse width of about 15 nsec and a pulse amplitude V_p of about 0.15 V_{DC} is found to be most suitable for the analysis of most alloys (see below). Due to the sharp drop of the field strength within a few nm above the tip surface, the field evaporated ions reach their terminal velocity v shortly after the desorption event and drift with an energy

$$n \cdot e(V_{DC} + V_p) = \frac{1}{2} mv^2 = \frac{1}{2} m \frac{d^2}{t^2} \tag{3.1a}$$

over a distance d (virtually the tip-to-detector distance; Fig. 3.1) until they strike the ion-detector after a time-of-flight t. In essence, Eq. (3.1a) already contains all measurable quantities for the evaluation of m/n; in practice, however, m/n is obtained from

$$m/n = 1927 \cdot (V_{DC} + \alpha V_p) \left(\frac{t' + \delta}{d} \right)^2 \tag{3.1b}$$

and is given in atomic mass units (a.m.u.) if V_{DC} and V_p are expressed in kV, and $t' + \delta$ and d in µsec and cm, respectively. Commonly, the real time-of-flight t is composed of the actually measured time t' and a constant electronic delay time δ; the pulse factor α (about 0.9) takes into account that the pulse amplitude experienced by the tip deviates from that generated at the high-voltage pulser, because of the improperly terminated pulse transmission line[85, 86]. Furthermore, the pulse shape is not ideally rectangular but rather shows some amplitude fluctuations during its duration[85]; therefore α may change – although slightly – not only with V_p, as is observed frequently, but also during the field evaporation process. This fluctuation in pulse amplitude determines to a large extent the mass resolution of the atom probe. In practice, α as well as the electronic delay time δ and the actual flight distance d (allowing for some small variations in specimen length) can be determined empirically for each FIM tip[86, 87].

An atom-probe cycle is initiated by the field evaporation pulse which triggers simultaneously an electronic timer (Fig. 3.1). After certain time periods t', the different channels of the timer receive stop signals from the ion detector in the sequence of the arriving ions. During each atom-probe cycle V_{DC} and V_p are also measured and hence, by knowing the charge state n (see Table 2.1), the mass m of each ion striking the detector can be evaluated by virtue of Eq. (3.1). Data acquisition and control of most atom-probes is performed by a computer system allowing several thousands of atoms per hour to be collected in most analyses. It might be possible, if the field evaporation rate is too high, that several ions of the same m/n ratio which have evaporated simultaneously arrive at the same instant at the detector. However, since the detectors used in atom probes do not discriminate the pulse height, each detector signal is attributed to the arrival of only *one* single ion. Therefore, the number of counted ions might be always less than the number of ions which actually arrive at the detector if more than one atom is field evaporated per pulse. In this case, in determining the composition of an alloy, the error becomes largest for the most abundant alloying species, i.e. the solvent atoms, giving too high a solute concentration. In order to avoid these problems, a fairly low field evaporation rate has to be chosen, i.e. between 10^{-1} and 10^{-3} ions per pulse.

3.2 Analysis of Atom-Probe Data

In principle, the microanalysis of alloys by means of the atom-probe FIM can be performed in several different ways, depending on the kind of information one is interested in.

In order to analyse the composition of a precipitated particle, which is discernible in the FIM, the field ion specimen is rotated by means of the manipulator (Chap. 2.2) until the image of the particular particle covers the probe hole. If the particle is smaller than the effective probe hole, the magnification has to be increased until it covers the probe hole entirely. The composition of the precipitate is now determined with the atom probe by collecting the atoms originating from a cylinder[1] whose diameter is given by the effective probe-hole size (in the order of a few nm) and whose length is confined by the total depth, which can be probed before the precipitate has been completely field evaporated (Fig. 2.18 a). By continuing to probe into the adjacent matrix and by determining the composition of each successively removed lattice plane, the concentration profile across interphase boundaries can also be obtained. Additionally, as has been discussed in

---

1 If the tip voltage is continuously raised during atom-probing the analysed volume is rather a truncated cone than a cylinder

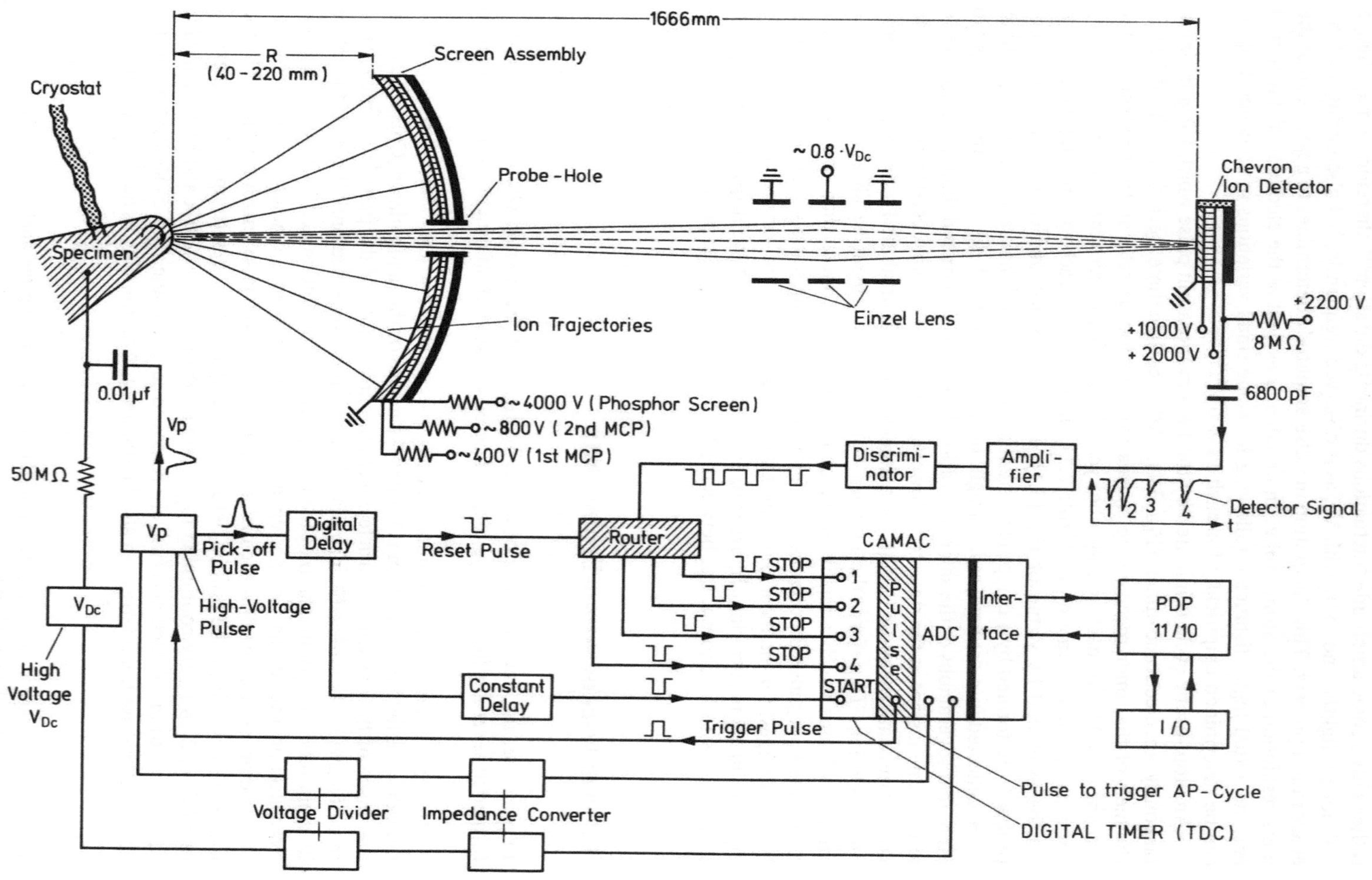

1666 mm
R (40 – 220 mm)
Cryostat
Screen Assembly
Probe-Hole
Specimen
Ion Trajectories
~ 0.8·V_Dc
Einzel Lens
Chevron Ion Detector
+2200 V
+1000 V
+2000 V
8 MΩ
6800 pF
~ 4000 V (Phosphor Screen)
~ 800 V (2nd MCP)
~ 400 V (1st MCP)
0.01 µf
Vp
Vp
50 MΩ
V_Dc
High Voltage V_Dc
Pick-off Pulse
High-Voltage Pulser
Digital Delay
Reset Pulse
Router
Discriminator
Amplifier
Detector Signal
1 2 3 4
t
STOP
STOP
STOP
STOP
START
1
2
3
4
Constant Delay
Trigger Pulse
CAMAC
Pulse
ADC
Interface
PDP 11/10
I/O
Pulse to trigger AP-Cycle
DIGITAL TIMER (TDC)
Voltage Divider
Impedance Converter

Fig. 3.1. Atom probe as designed by the author and operated at the University of Göttingen. The specimen is mounted on an x-y-z manipulator and attached to a glass-insulated cryostat (filled with either liquid N_2 or liquid H_2) via a flexible copper braid. The tip-to-screen distance (R) can be varied between 40 and 180 mm by shifting the screen assembly. The latter is either used as FIM screen or as detector for the imaging atom probe (IAP). The screen assembly is curved (75 mm in diameter; radius of curvature 120 mm) in order to provide equal flight distances in the IAP mode; the diameter of the probe-hole, which is internally grounded by a metal tube to avoid electrostatic deflections of the ion beam, is 2.5 mm. To maximize the ion yield per field evaporated plane the specimen is aligned onto the instrument axis through x-y-z manipulation until the current caused by randomly arriving imaging gas ions is at a maximum; the spread of the ion beam which is imaged on the Chevron detector, is focused by adjusting the potential at the Einzel lens. Atom probe control and data acquisition are performed by means of a pdp-11/10 computer which is interfaced to a CAMAC-system consisting of 1) a 4-channel 200 MHz clock (TDC) to measure the ToF; the 4 channels of the clock are simultaneously started by a pulse which is synchronously picked-off from the high-voltage pulser. 2) A pulse module providing two pulses: i) for triggering the Krytron tube-operated high-voltage pulser and, thus, for initiating an atom-probe cycle, and ii) to reset the router. If a sequence of ions (up to four) rather than a single ion strikes the detector, the router assigns a stop signal from the first ion of the sequence to channel 1 of the TDC, the second one to channel 2, etc. 3) The AD converter measures V_P and V_{DC} during each atom probe cycle.

The vaccum system (part of which is shown in Fig. 2.3) consists basically of the specimen exchange chamber and the main chamber, which contains the FIM; both are separated by a straight-through valve. For easy and rapid specimen exchange without spoiling the vacuum in the main chamber, 11 specimens can be stored on a revolving plate in the specimen chamber. With respect to the vacuum system, the only provision made to incorporate the ToF spectrometer into the FIM is a flight-tube with a detector housing at its end. For the sake of simplicity the electronic circuitry for the operation of the time-gated imaging atom-probe, which is integrated into the same system, is not included but is shown separately in Fig. 3.6. (For more details see Ref. 7)

Chap. 2.7.1, the morphology of the particular precipitate can be determined accurately during the course of the analysis. This type of analysis for the determination of the composition of single particles is essentially based on the visibility of the precipitate in the FIM image and its accuracy is mainly governed by the total number of atoms which can be collected from the precipitate particle; if the particle is small it is necessary to analyse several particles. This, of course, is only possible if the volume density is sufficiently high.

In order to analyse the composition and distribution of particles which are not visible in the FIM (Chap. 2.6.3) or in order to obtain information about solute fluctuations in alloys (e.g. spinodally decomposing alloys), it is necessary to determine the concentration profile in a cylinder of considerable length (e.g. more than 100 nm along a particular direction [h'k'l']) with an extremely good depth resolution (Fig. 2.18b). For this purpose, the planar concentration of each successively field evaporated (h'k'l')-lattice plane is determined and plotted against the number of evaporated planes; hence, the depth resolution is equal to the interplanar spacing d(h'k'l'). Whenever the analysed cylinder cuts completely through a particle (Fig. 2.18b), its composition and the solute variation across its interphase boundary can be determined directly from the concentration profile. However, if the particle is cut partly by the analysed cylinder (Fig. 2.18b), the concentration profile only reveals a local increase in the solute content as compared to the surrounding matrix but does not give any information about the true composition of the particle. Figure 3.2 shows the concentration profile along [111] of an aged Cu-Ti-alloy, which contains discrete precipitates having the stoichiometric composition Cu_4Ti, e.g. the particles contain 20 at% Ti. It is recognized that three Cu_4Ti-particles are completely cut (A) and that two precipitates are probably only partly cut (B) by the analysed cylinder.

From Fig. 3.2 it may also be noticed that there is considerable statistical noise superposed on the real concentration profile. This statistical noise is mainly due to both the rather poor detector efficiency (approximately 60% for the most frequently used Chevron-detector[88]) and the limited number N of atoms which can be collected per field evaporated plane. (N depends on both the effective probe-hole size d_{ap} and the areal density of atoms in this plane; however, during a sequence of successively field evapo-

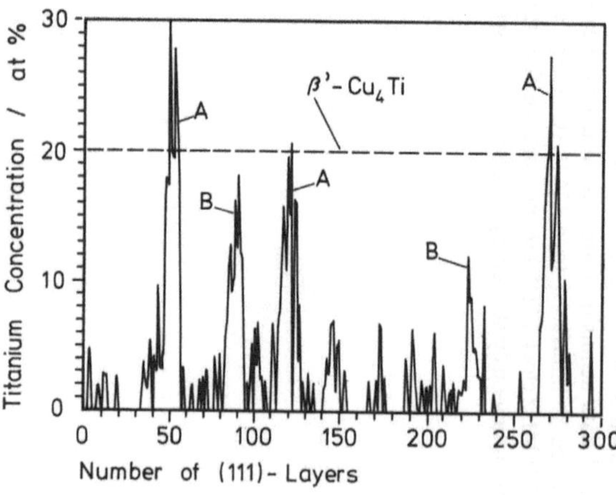

Fig 3.2. Concentration profile along [111] of a Cu-2.7 at% Ti specimen aged for 500 min at 350 °C. In the average, 50 atoms per field evaporated (111) plane were collected. One removed (111)-layer corresponds to a probing depth of 0.18 nm. Particles completely and only partly cut by the analysed cylinder are denoted by A and B, respectively[89]

rated planes, N is found to fluctuate by up to approximately 30% about a mean value $\overline{N}$, which ranges typically between 50 and 120 for $d_{ap} \approx 2$ to 3 nm.)

In order to account for the statistical significance of atom-probe data, it is in general necessary to place confidence limits on each data point. In analysing a solid solution of A and B atoms by recording concentration profiles, the entire sampling volume is usually subdivided into n_0 small subvolumes V_i, each of which is considered to be homogeneous; the size of V_i is governed by both the required depth resolution and the necessary accuracy of the analysis; its size is irrelevant when the solid solution is random. However, for the detection of small deviations from randomness in the composition profiles, e.g. in the presence of solute clusters, V_i can not be chosen much larger than the mean cluster size. From each of these subvolumes $N^* = N_A^* + N_B^*$ atoms are collected allowing the concentrations $c_A^* = N_A^*/N^*$ and $c_B^* = N_B^*/N^*$ of A and B atoms to be calculated in this particular volume. The measured values c_A^* and c_B^*, which must be considered to approximate best the true concentrations c_A and c_B in each subvolume, are expected to be binomially distributed; hence, the standard deviation (σ) of c_A^* is simply given by

$$\sigma = (c_A(1 - c_A)/N^*)^{1/2} \approx (c_A^*(1 - c_A^*)/N^*)^{1/2} . \tag{3.2}$$

It is immediately evident from Eq. (3.2) that small confidence limits require N^* to be rather large. This, for instance, can easily be achieved in a composition analysis of large precipitates, where V_i may be chosen to be almost equal to the volume of the precipitated particle. In general it is, however, not possible for the analysis of composition fluctuations in alloys, if ultimate depth resolution, i.e. one interplanar spacing d_{hkl}, has to be achieved, since commonly only less than about 100 atoms can be collected from the associated subvolume $V_i = (d_{ap}/2)^2\pi \cdot d_{hkl}$ (see above); therefore, composition profiles such as shown in Fig. 3.2 are blurred by rather large statistical fluctuations, which impede a straight-forward interpretation of the true concentration of a localized solute fluctuation.

If a smaller depth resolution can be tolerated, smoothing of the composition profiles can be achieved in two different ways: i) V_i and, hence, N^* is chosen larger by summing up the atoms from several planes or ii) by performing a moving-average analysis. In this sort of smoothing technique n successive planes form a window. This window is placed onto the first n planes of the original composition profile and the average composition within the window is determined with $\sigma = (c_A^* c_B^*/N_n^*)$, where N_n^* denotes the total number of atoms collected from the n planes covered momentarily by the window; subsequently the window is moved one interplanar distance and again the average composition is determined, etc. This *smoothing technique* often allows – if n is properly chosen – the deconvolution of small solute clusters, which have not been unambiguously discernible in the original, blurred composition profile. It must, however, be emphasized that these smoothing techniques, in fact, remove the high frequency components of the original composition profile, but also will flatten the concentration profile across interphase boundaries. For this reason the original composition profile with its ultimate depth resolution and poor statistics ought to be carefully inspected prior to applying any smoothing procedure.

By analogy with the analysis of time series, the data forming the concentration profile can also be analysed in terms of an *autocorrelation analysis,* provided the probed distance through the specimen is sufficiently large. The correlation coefficient R(k), given by

$$R(k) = \frac{n}{n-k} \cdot \frac{\displaystyle\sum_{i=0}^{n-k} (c_i - c_0)(c_{i+k} - c_0)}{\displaystyle\sum_{i=1}^{n} (c_i - c_0)^2} , \qquad (3.2)$$

decays towards k_0 (Fig. 3.3), which corresponds to the length of positive correlation. (c_i and c_{i+k} denote the solute concentration in the i-th and (i + k)-th field evaporated plane (h'k'l'), respectively; c_0 is the mean solute concentration and n is the total number of field evaporated planes; hence, the correlation distance k is also expressed in terms of the number of successively removed (h'k'l')-layers.) The average diameter $\overline{D}$ of non-statistical deviations from the mean in the concentration profile, such as for solute-rich clusters or discrete precipitates, is now simply given by $\overline{D} = d(h'k'l') \cdot k_0$[89, 90]. Furthermore, the correlogram also contains information about the precipitated volume fraction[90] and the distribution of interparticle spacings. In Fig. 3.3, the occurrence of the first peak at k_1 (corresponding to 35 removed (111)-layers) indicates a quasi-periodical distribution of precipitates with an interparticle distance $\lambda = k_1 \cdot d(h'k'l')$. The second peak at k_2 occurs at exactly twice the distance (70 removed (111)-layers) of the first peak confirming the quasi-periodicity of the particle arrangement. For randomly distributed solute-rich regions of small volume density R(k) remains zero for $k > k_0$.

Additional information about the periodicity and waveform of long-range composition fluctuations, e.g. as they are expected to form during the early stages of spinodal decomposition[89, 92], can be obtained by *Fourier transforming* the concentration profile. It

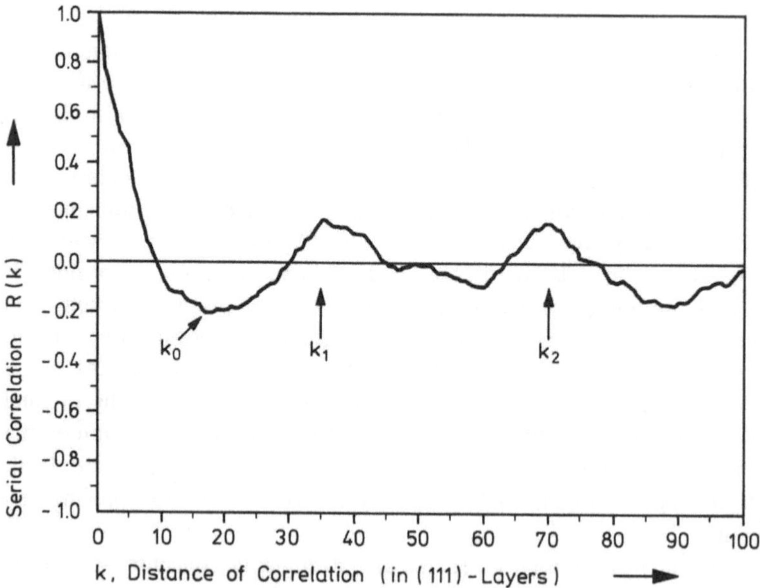

Fig. 3.3. Correlogram of the concentration profile displayed in Fig. 3.2. See text

has been shown[89] that the Fourier spectra of aged Cu-Ti-alloys consist of several Fourier components; however, the frequency of the main Fourier component agrees always quite accurately with the interparticle distance (k_1) as determined from the corresponding autocorrelation analysis.

As was discussed above, the limited number of atoms per field evaporated plane introduces a considerable statistical error into the determination of the planar concentration. In order to verify to what extent the concentration profiles, as determined by the atom probe, are relevant to the decomposition state or whether they are essentially obscured by statistical noise, *computer simulations* of the atom-probe experiment have to be performed. For this purpose, various concentration profiles, which might represent the actual concentration profile in the specimen, are subjected to simulated atom-probe analyses with varying experimental parameters, such as size of the probe hole, number of atoms per plane, and crystallographic direction into which it is probed[89]. The various simulated "noisy" profiles as well as the corresponding simulated Fourier spectra and correlograms can then be compared with the corresponding presentation of the experimental data.

Figure 3.4 a shows a heterogenous alloy consisting of three different phases α (0.5 at% solute concentration), α' (2.7 at%), and β (25 at%), which was subjected to a computer-simulated atom-probe analysis assuming a mean detection efficiency of $\overline{N} = 50$

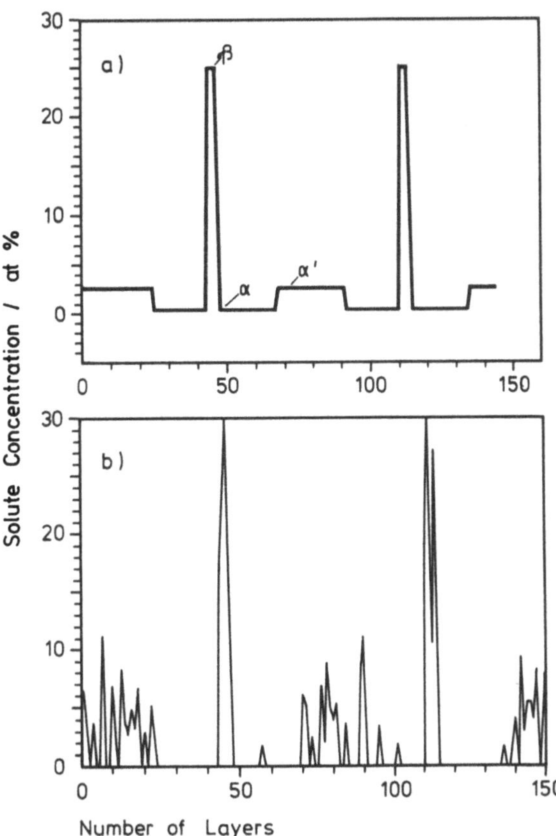

Fig. 3.4 a. Part of an ideal concentration profile from a heterogeneous alloy consisting of three different phases α (depleted matrix), α' (supersaturated matrix), and precipitated phase β. In **b**) this alloy was subjected to computer simulations assuming a collection efficiency of $\overline{N} = 50$ atoms per atomic plane. Although the statistical fluctuations are rather large the original profile, shown in **a**), is still to be recognized[89]

atoms per field evaporated plane. Although the statistical fluctuations caused by the small sampling of only 50 atoms blur the original concentration profile significantly, the three individual phases of the decomposed alloy can still be qualitatively recognized in Fig. 3.4b. By comparison of the original profile (Fig. 3.4a) with the simulated one (3.4b), the error involved in this type of atom-probe analysis can also be estimated.

In conclusion, a combination of various analyses of atom-probe data together with the computer simulations yield the most comprehensive information about the microstructure of decomposing alloys.

3.3 Problems and Limitations of Atom-Probe Analyses

We have seen in the previous chapters, that the rather poor statistics are often the main hindrance to analyse a concentration profile in an alloy with the optimum depth resolution achievable in the atom probe, i.e. one interplanar spacing. In fact, the statistics could be improved by increasing the effective size of the probe hole; however, the maximum size of the probe hole is confined by the spatial extension of solute fluctuations or by the diameter of precipitates and, since the most powerful application of the atom-probe FIM is the analysis of small clusters, the probe hole has to be kept fairly small (2 to 5 nm) in most experiments. Hence, the true achievable depth resolution is mainly governed by the microstructural parameters themselves and may vary from experiment to experiment; the optimum probe-hole size can often be determined by computer simulations (Chap. 3.2).

In principle, the lateral spatial resolution is controlled by the effective probe-hole size as long as the specimen end-form is regular; in this case a lateral resolution of about one atomic diameter, i.e. 0.2 nm, is obtainable as has been demonstrated by Krishnaswamy et al.[93] by aiming for single rhenium atoms which show a bright spot contrast in a tungsten matrix (Chap. 2.6.1). However, in more realistic cases, where the local radius of curvature or/and the field evaporation field varies sharply as, e.g., for small precipitates or grain boundaries, often the trajectories of ions from the precipitates or grain boundary region and the adjacent matrix may overlap[94]. In these unfavourable circumstances, the best lateral resolution and, hence, the smallest diameter of particles, which can still be analysed quantitatively, may be limited to 5 nm as was pointed out by Miller et al.[84].

In analysing the composition of a particular phase in an alloy composed of two or more elements, one is in general faced with several problems. i) As has been discussed in Chap. 2.6.1 the evaporation conditions for the various species forming the alloy can be different, i.e. some species evaporate already at the imaging voltage V_{DC} between two successive field evaporation pulses and, hence, are not recorded for the analysis, while others are only removed when the high-voltage pulse is applied. In this case, the concentration of the atoms which are more resistant against field evaporation will turn out to be too high. To prevent the *selective removal* of any species between the desorption pulses, V_{DC} has to be lowered whereas the pulse amplitude V_p has to be increased by the same amount until all species will evaporate only when the pulse is applied. However, since V_{DC} has to be sufficiently high in order to avoid corrosion of the specimen by adsorption of residual gas atoms, the pulse fraction V_p/V_{DC} may not be chosen too high. The appropriate pulse fraction often has to be determined empirically by checking whether the composition of the phase under investigation is dependent on the choice of V_p/V_{DC}; several studies have reported an optimum pulse fraction between 0.1 and 0.15[7, 84]. ii) There are cases where two isotopes of different chemical elements have the same specific masses m/n; e.g., in an alloy containing Ti and Mo the most abundant Ti-isotope (48 a.m.u), which evaporates from most alloys as doubly charged ions, overlaps on the m/n-scale at 24 a.m.u with the 4-fold charged Mo-isotop of 96 a.m.u. Fortunately, there exist in most cases some isotopes of the same elements which do not overlap[84]. By counting their numbers and by using tables of the natural abundance of isotopes, it becomes possible to evaluate the corresponding number of ions hidden in the peak of the overlapping isotopes. iii) This procedure requires that the atom probe is capable to resolve the single isotopes of most alloy-forming elements, i.e. the mass resolution has to be of the order $\Delta m/m = 1/150$ (related to the measured peak width at 50% of peak height) under typical operating conditions; this value is actually achieved in most instruments. However, it has been observed in many quantitative analyses, that the mass peaks in a linear mass spectrum consist of a sharp leading edge with a frequently gradually decaying tail, the latter resulting from energy deficits of the field evaporated ions (Chap. 3.1) or from the formation of $(\text{metal-hydrogen})^{n+}$-ions of somewhat larger specific masses than the corresponding $(\text{metal})^{n+}$-ions. Sometimes the tail of one peak interferes with the peak of an element (or isotope), which is adjacent on the m/n-axis of the spectrum; in these cases the mass resolution is somewhat deteriorated. In order to deconvolute adjacent mass peaks, Miller et al.[84] proposed plotting the mass spectra semi-logarithmically. This plot yields a linear trailing edge which can be used to deconvolute adjacent isotope peaks. For this purpose, a series of straight lines, which are parallel to the trailing edge of the last peak of the overlapping isotopes, is drawn for each trailing edge of the preceding peaks; the area between these lines thus provides an empirical measure of the number of ions in each of the overlapping peaks[84]. iv) In order to obtain a detailed picture of solute-rich clusters with sufficient statistical reliability from concentration profiles and subsequent autocorrelation analyses (Chap. 3.2), the cluster density must be larger than approximately 10^{16} cm^{-3}, indicating that this type of atom-probe analysis is mainly confined to early stage decomposition reactions, i.e. before large scale coarsening of precipitated particles has occurred. In later coarsening stages one often has to rely on the analysis of one or two (visible) large precipitates accidentally found in the tip surface; furthermore, the analysis of precipitated phases of low volume density requires always complementary transmission-electron-microscopic investigations. v) The reliable minimum level of quantitative elemental detection depends strictly on both the background noise level and the number of ions collected during the analysis and, therefore, on the size of the precipitates or the depth of the concentration profile. In best practical cases, the minimum concentration, which can be determined quantitatively, is approximately 0.2 at%[84], although most elements can be detected qualitatively at a lower level.

In principal, all elements being constituents of a metallic alloy can be identified in the atom probe including the light elements such as Be, B, C, N, O, which are commonly not detectable by means of energy-dispersive X-ray analyses (EDX). Furthermore, with the above given limitations in mind, atom-probe analyses do not rely on either calibration standards or knowledge of ionization cross sections nor on complex background corrections, which often obscure the results obtained from various other microanalytical techniques.

3.4 The High-Resolution Energy-Focusing Atom Probe

The energy deficits of the field evaporated ions limit essentially the mass resolution of the conventional *straight atom probe*[95]. Hence, the mass resolution of the atom probe can be improved significantly by employing an energy focusing ToF mass spectrometer in the Poschenrieder configuration[96]. Such a *high resolution atom probe*[95] consists of a combination of a straight drift tube with a 163° toroidal deflector system (Fig. 3.5). Ions which are field evaporated simultaneously within the time period Δt which the electronic timer is capable of resolving, i.e. approximately 5 ns, but which experience different acceleration potentials because of changing pulse amplitude, are forced by the deflectors to strike the ion detector simultaneously within Δt. Thus, the mass resolution of this instrument is mainly controlled by the time resolution of the electronic timer; for such an instrument $\Delta m/m$ has been shown to be of the order of 1/5000, based on the measured peak width at half-maximum height[95].

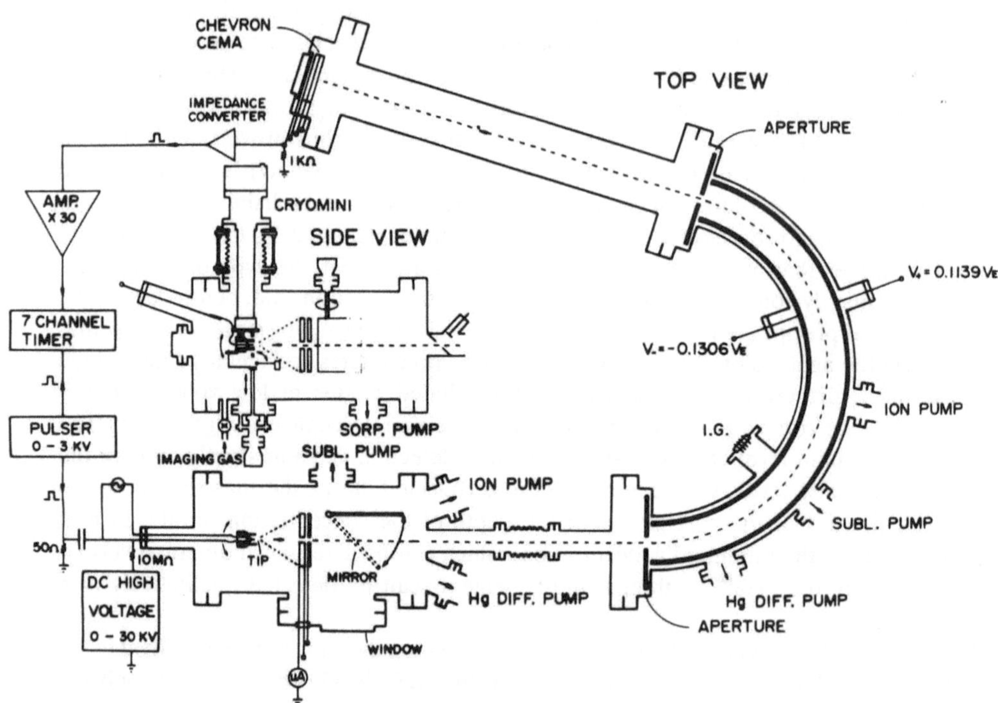

Fig. 3.5. The experimental arrangement of a high resolution energy focusing atom probe with 163° toroidal deflector plates (After Nishikawa et al.[97])

Although the mass resolution of the high-resolution energy-focusing atom probe is considerably higher than that of the straight atom probe, the performance of the latter is sufficient for most metallurgical applications.

3.5 The Field-Desorption Microscope and the Imaging Atom Probe

The principles and, hence, the operation of the field-desorption microscope (FDM) are quite similar to that of a field-ion microscope. However, unlike the FIM, in which the image is formed by gas atoms ionized above the high-field regions of the tip surface, the field-desorption microscope requires no ambient image gas atoms but the ions field evaporated or field desorbed (in the case of weakly bound adsorbed surface species, see Chap. 2.4) from the surface are used themselves to form the so-called field-desorption image[98]. A multilayer desorption image, such as is shown in Fig. 2.8 for aluminium, is obtained by either field evaporating continuously or by applying successively many evaporation pulses to the tip while the shutter of the recording camera remains open. Thus, the desorption image is composed of ions of different m/n-values (e.g. Al^{1+} and Al^{2+} in Fig. 2.8) evaporated or desorbed from the surface at the particular applied voltage; it displays the same crystallographic information as the corresponding field-ion image which is often formed prior to the desorption image in order to establish the proper voltage settings for field evaporation. It is impossible to distinguish in the desorption microscope between species of different specific masses m/n and, therefore, it is not possible to determine the distribution of different elements in the surface of a tip which is an alloy rather than a pure metal.

In order to achieve a distinction and, hence, an identification of different elements, the desorption microscope is operated as an imaging atom probe (IAP)[4-7], which is schematically shown in Fig. 3.6 (a comprehensive review about the instrumental design was given recently by Panitz[99]). As was discussed in the context of the ToF atom probe (Chap. 3.1) the time-of-flight of the field evaporated ions is dependent on their specific mass m/n. Therefore, ions which are simultaneously released from the surface in the instant when the high-voltage pulse is applied and which have a different m/n, will strike the detector of the imaging atom probe after different times-of-flight, which can be calculated from Eq. (3.1) for any given m/n, or measured directly from the spectrum of ions which arrive after one single evaporation pulse on the detector of the field desorption microscope. (In order to obtain this spectrum, the detector current is monitored as a function of time on an oscillosope.) The detector of the imaging atom probe which consists of a sandwich of one or two spherically curved image-intensifying microchannel plates and a phosphor screen may be switched on by electrically-delayed high-voltage pulses of about 15 ns duration; after 15 ns the detector is again switched off. By calculating the time-of-flight of a preselected ion species (Eq. (3.1)) and by choosing the proper delay time, the on-time of the detector can be brought into coincidence with the arrival of these preselected ions. Thus, the image on the phosphor screen, which is recorded photographically, then reveals the distribution of their arrival positions. Figure 3.7 shows the IAP-images of an aged Al-2 at% Cu tip; each bright spot indicates the arrival position of one ion. The left part of Fig. 3.7 was obtained by gating the detector of the IAP

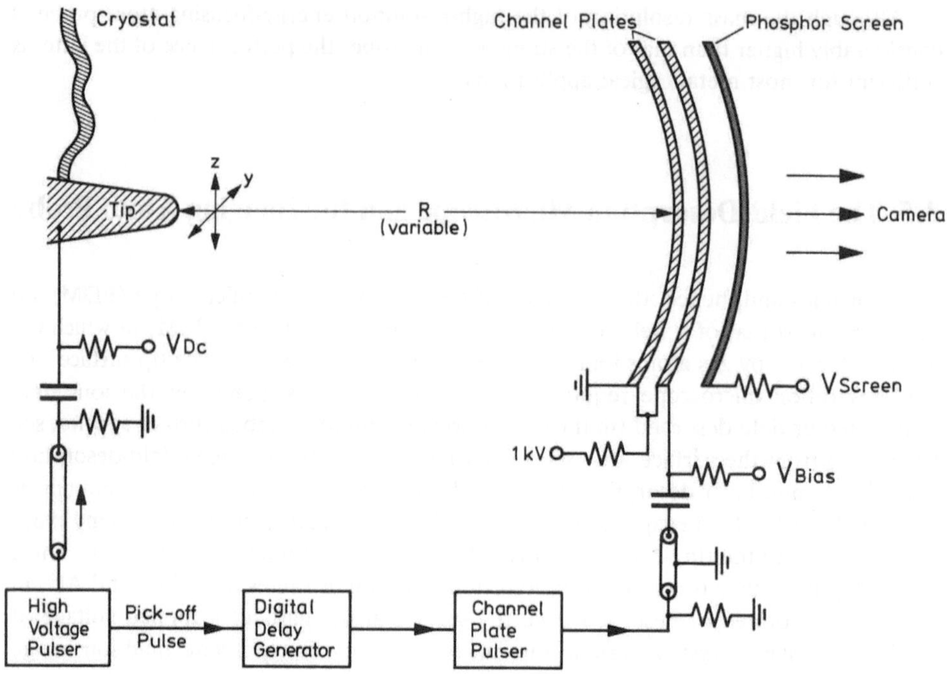

Fig. 3.6. Basic design and electronic circuitry of an imaging atom probe. See text and also Fig. 3.1

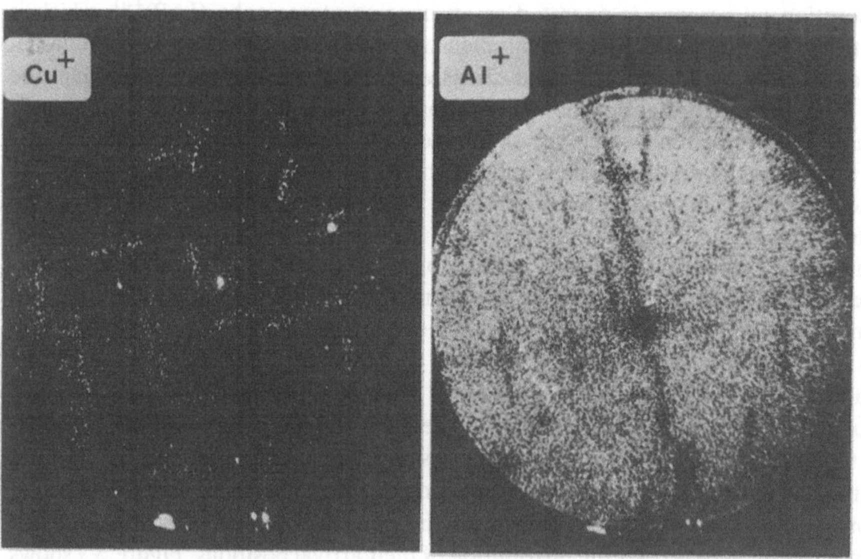

Fig. 3.7. IAP images of an aged Al-2 at% Cu alloy. Gated for Cu^+ ions (*left*); gated for Al^+ ions (*right*). (Courtesy A. R. Waugh)

for Cu^+ ions (m/n = 63); the enrichment of Cu in the precipitate zones is clearly to be seen. In the right part of Fig. 3.7, the detector was gated for Al^+ ions (m/n = 27) and, consequently, the Cu-rich zones remain dark.

In this context, it should be emphasized that after field evaporation of a particular atom from its surroundings, the electric field in its original vicinity may be altered by electronic and/or geometrical factors, leading to an undefined change in the trajectories of the subsequently field evaporated atoms. Therefore, adjacent bright spots in an IAP image, or two ions successively reordered with the atom-probe FIM need not necessarily have been next neighbours in the lattice prior to evaporation. As a consequence, analyses of atomic arrangements on an atomic scale (e.g. short-range ordering) or atomic relaxations are not feasible with either the IAP or the atom-probe FIM.

As we have discussed in the previous sections, the conventional ToF atom probe is most powerful in providing quantitative analyses of small precipitates and long-range compositional fluctuations but much information contained in the specimen regions lying outside the small probed area is lost. It is the advantage of the time-gated IAP to yield information about the spatial distribution of preselected species over the entire part of the specimen surface, which is imaged in the FIM. (Usually, the FIM images, which often are recorded prior to the IAP images, are compared with the latter or even superimposed [4].) Despite the rather poor mass resolution ($\Delta m/m$ = 1/20 to 1/50) and the fact, that most IAP analyses are only qualitative in the sense that only the spatial distribution of one ion species is obtained (and, hence, no local concentration can be determined), the IAP is very useful for the detection of sharp concentration gradients across interphase boundaries and grain boundaries[4]; it provides also the means to image phase heterogeneities, which do not show any contrast in the field ion image (Chap. 2.6.3). Hitherto, the IAP has not yet been applied to its full potential, but it seems that for most metallurgical studies the range of applicability of the ToF atom probe is broader than that of the IAP. However, since it is possible to integrate an IAP into an atom-probe FIM[100], both techniques can easily be applied simultaneously to the investigation of the same specimen.

4 Applications and Results

4.1 Overview of Defect Analyses by Means of FIM and Atom Probe

In the following sections we shall discuss the application of quantitative field-ion spectroscopy to studies of the microstructure of metals and alloys. We shall confine ourselves to both the discussion of results from only more recent investigations, which have contributed significantly to the understanding of the kind and development of defect structures in those materials, and to the discussion of some controversial results.

Table 4.1 summarizes the metallurgical problems and the related defects studied so far by means of either FIM, atom-probe FIM, imaging atom probe or a combination of these techniques. As indicated in Table 4.1, quantitative field-ion spectroscopy has been applied in the past to the investigation of all defects which are relevant in metallurgy and materials science. Significant information which probably could not have been obtained

Table 4.1. Metallurgical problems and related defects studied by FIM, atom probe FIM, or imaging atom probe

Dimension of defect	Defects	Problems	Applied technique
0	vacancies, interstitial impurities self-interstitials	distribution, density annealing of vacancies annihilation of vacancies and self-interstitials	FIM, imaging atom probe
	solute atoms	short-range ordering and clustering	FIM
1	dislocations	characterization, core structure	FIM
2	grain boundaries	structure, segregation	FIM, imaging atom probe
	interphase boundaries	structure, concentration gradient across boundary	FIM, atom probe FIM, imaging atom probe
3	second phase precipitation,	characterization of phase transformation, morphology, size, distribution, composition, growth kinetics	atom-probe FIM, FIM (imaging atom probe)
	vacancy and impurity clusters, voids, depleted zones	formation, morphology, size, annealing kinetics	FIM, atom probe FIM

by any other microanalytical techniques, has been gained about defects of zero, two and three dimensions. However, the application of the FIM to the analysis of line defects has to be considered with some reservations. Although, there exists a comprehensive contrast theory for line defects (e.g. Ref. 10), the FIM has not yielded much information about the arrangement of atoms in the dislocation core, mainly because of insufficient resolution and the strong forces acting on the dislocation during imaging. As outlined in Chap. 2.5 the operative high stresses only allow observation of dislocations in refractory metals; in non-refractory metals the presence of glissile dislocations leads to immediate rupture of the tip when the electric field is applied. Moreover, because of the small imaged area, only single perfect dislocations, small dislocation loops, and stacking faults bound by not too widely separated partial dislocations can be identified; studies of dislocation networks (other than structural networks found in some grain boundaries, see Chap. 4.5.1) and dislocation-dislocation interactions, however, are almost impossible. It should also be noted that the observed dissociation width of a dislocation is influenced by both the field-induced shear stresses and the image forces operative at the surface. For example, the dissociation width of a dislocation in a bcc metal is known to be the order of 0.5 nm, whereas FIM studies revealed about 10 nm[10]. As a result of these problems, the interest in analysing line defects in the FIM ceased simultaneously with the development of TEM high-resolution techniques, such as weak-beam[101] and direct lattice imaging[102], which yield much more detailed and more reliable information about line defects and the dissociation of dislocations into partials.

4.2 Point Defects

4.2.1 Some General Remarks

The types of analysed individual point defects may be divided into two categories: i) defects, which are constituents of the metal in thermodynamic equilibrium, such as thermal vacancies, chemical vacancies in some intermetallic compounds of non-perfect stoichiometry, and substitutional as well as interstitial impurities; ii) extrinsic point defects, such as vacancies and self-interstitials, produced by irradiation with high-energy particles or by ion implantation, which are not in thermal equilibrium. Although it is obvious, that the FIM with its atomic resolution is most suited for the study of the above mentioned point defects (and often is the only direct method) the analyses are by no means straightforward because of difficulties in contrast interpretation and the danger of artifacts as has been outlined in Chap. 2.6.1 and 2.6.4. Furthermore, the volume sampled in the FIM is very small (in the order of 10^{-21} m^3) by comparison with other microanalytical techniques; this fact, inherent to field-ion microscopy, sometimes affects the reliability of the quantitative determination of point defect concentrations.

4.2.2 Studies of Thermal Vacancies

The only systematic and reliable investigation of thermal vacancies by means of FIM was performed by Berger et al.[63] in platinum quenched from 1700 °C. Control experiments on well-annealed pure Pt specimens yielded an abnormally high artifact monovacancy concentration of up to 1.4 at% if black spot counting (Chap. 2.6.4) was performed on either {012} or {137} planes, whereas the artifact monovacancy concentration on various other planes (e.g. {135}, {124}, {167}) remained below 4×10^{-5} at.fr.; therefore, the {012} and {137} planes were disregarded for the study of thermally generated vacancies in quenched specimens. In the as-quenched specimens Berger et al. identified 157 monovacancies and 9 divacancies in a total of 593 794 atomic sites which were scanned; besides the few divacancies, no multiple vacancies were detected. This result indicates that the monovacancy is the dominant defect in thermal equilibrium even at elevated temperatures. After correcting for the background artifact monovacancies, the ratio of divacancy and monovacancy concentrations was 0.06 ± 0.02 and the quenched in monovacancy concentration was $(2.64 \pm 0.14) \cdot 10^{-4}$ at.fr. This value is considerably lower than the caluclated monovacancy concentration at 1700 °C, indicating that defect losses occur during the quench in the elevated temperature regime where the vacancy mobility is sufficiently high. However, at a critical temperature T^+, the defects become frozen in; hence, the determined defect structure corresponds to that at T^+, which was evaluated by Berger et al.[63] to be 443 °C. Based on these experimental data and the knowledge of the monovacancy migration energy together with an approximate value for the frequency factor for vacancy migration, the divacancy binding free energy at $T^+ = 443$ °C was calculated to be 0.23 eV. By partially annealing the as-quenched specimen at around 350 °C, the concentration of divacancies decreased by a factor of ~ 5, whereas the monovacancy concentration decreased only by approximately 16%. This result gives one of the few direct pieces of evidence that a divacancy is more mobile than a monovacancy. This study has also clearly demonstrated that it is possible, although quite tedious, to obtain statistically meaningful results about quenched-in vacancy defects of relatively small concentrations by means of FIM.

In a recent study, Paris et al.[103] tried to determine the distribution and concentration of vacancies in an ordered Fe-49.5 at% Al. For this purpose the alloy was quenched from 1000 °C; the control specimen was subsequently annealed for 24 h at 425 °C. In FIM images of the ordered Fe-Al alloy only Fe atoms are visible and, therefore, only sites belonging to the Fe sublattice can be analysed; the Al atoms remain dark either because of insufficient field ionization above them or because of selective field evaporation (Chap. 2.6.1 and 2.6.2). Therefore, it is impossible to decide whether a dark site in the Fe sublattice is due to a vacancy or whether it is caused by imperfect long-range ordering, i.e. by a misplaced aluminium atom sitting on the iron sublattice. Although the extremely large vacancy concentration ($\sim 2.5 \cdot 10^{-2}$ at.fr.) observed in well annealed control specimens points towards the latter contrast interpretation, Paris et al.[103] based their analysis on the assumption that all detected dark spots are associated with vacant sites in the Fe sublattice. Because of these unresolved contrast problems, and also because of the statistically insufficient number of only 2500 examined sites, the evaluated concentration ($1.2 \cdot 10^{-2}$ at.fr.) of quenched-in vacancies on the Fe sublattice must be regarded with caution.

4.2.3 Interstitial Impurities

In a number of investigations it has been shown that interstitial oxygen in Pt, Ir and W gives rise to bright spots (Chap. 2.6.4[104–106]). However, as in the case of substitutional impurities (Chap. 2.6.1) and vacancies, there exists again no generally valid one to one correspondence between the number of bright spots and the interstitial oxygen concentration. For instance, in W and Ir it is necessary to count bright spots away from the [001]-zone decoration line[105, 106] in order to avoid counting of artifact bright spots. By confining the counting to {111} regions, Fortes and Ralph[105] determined an oxygen content of 470 ppm in a Pt tip whereas chemical analysis yielded 500 ppm. (The segregation of interstitial oxygen to grain boundaries will be discussed in Chap. 4.5.3.)

Papazian[107] has shown, that additions of carbon (0.2 w%) or nitrogen (up to 0.05 w%) to a Fe-4 w% Mo alloy do not change the original imaging characteristics. This result supports the conclusion from an earlier investigation of carbon in tungsten by Machlin[106] that carbon atoms remain invisible in the FIM.

4.2.4 Short-Range Order and Clustering

As was pointed out in Chap. 2.6.1, the sites of gold and nickel atoms in binary platinum based solid solutions, containing small concentrations of these elements, can be identified in the FIM by their appearance as α-dark spots. Repeating a previous study by Gold and Machlin[108, 109], who did not differentiate between the real solute sites, i.e. between α-dark spots and the artificial β-dark spots (Chap. 2.6.1), Chen and Balluffi[58] succeeded in determining the exact positions of large numbers of solute atoms in the three-dimensional lattice of Pt-4 at% Au, Pt-0.62 at% Au, and Pt-0.65 at% Ni by pulsed field evaporation. These data have been used to generate the normalized radial distribution function

$$R(r_i)/R_{ran}(r_i) = \frac{2 N_{AAi}}{\overline{Z}_i m_A N_A} .$$
(4.1)

Here $R(r_i) = 2 N_{AAi}/N_A$ denotes the average number of solute atoms around a solute atom at the distance r_i of the i-th nearest neighbour shell; N_{AAi} is the number of i-th nearest-neighbours solute atom pairs. $R_{ran}(r_i)$ is the corresponding radial distribution function in a solid solution exhibiting neither short-range ordering nor clustering tendency; $R_{ran}(r_i)$ is simply expressed as $\overline{Z}_i \cdot m_A$, where $m_A = N_A/N$ is the fraction of solute atoms. Since the coordination numbers of atoms near the surface are reduced as compared to atoms in an infinite crystal, the average effective coordination numbers $\overline{Z}_i$ had to be calculated from the experimental data for each analysed volume.

In Fig. 4.1 the normalized radial distribution functions according to Eq. (4.1) are shown for the first 9 nearest neighbours. For Pt-4 at% Au, the ratio $R(r_i)/R_{ran}(r_i)$ is greater than unity for all measured distances r_i; thus, the solid solution exhibits a clustering tendency. In contrast, in a Pt-0.65 at% Ni alloy $R(r_i)/R_{ran}(r_i)$ oscillates around R/R_{ran} = 1, indicating short-range ordering. As was pointed out by Chen and Balluffi (and recently by Machlin[110]) these results are in agreement with thermodynamic measurements.

Once the radial distribution function has been generated, the Warren short-range order coefficients α_i may be calculated by virtue of

$$\alpha_i = \left(\frac{R(r_i)}{R_{ran}(r_i)} - 1 \right) \frac{m_A}{m_B} .$$
(4.2)

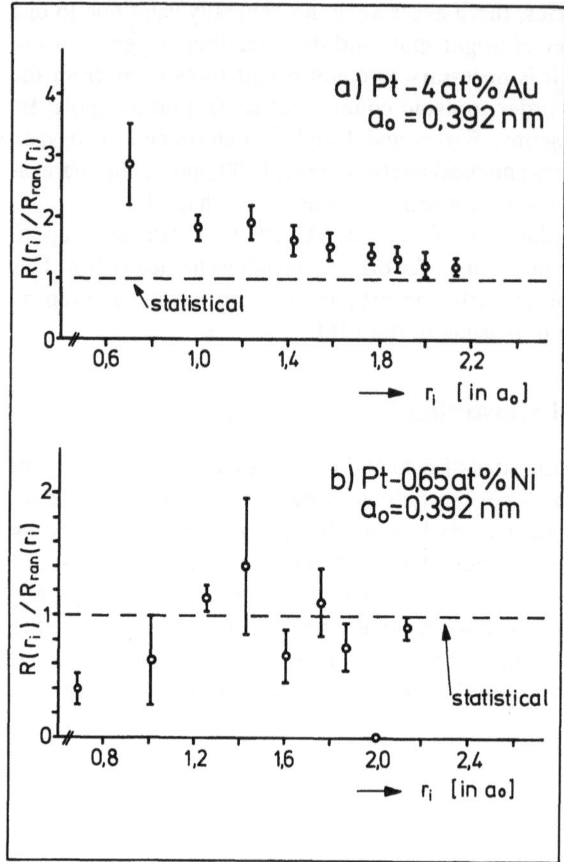

Fig. 4.1a, b. Radial distribution function of gold atoms in a Pt-4 at% Au alloy (a) and of nickel atoms in a Pt-0.65 at% Ni alloy (b). (After C. G. Chen and R. W. Balluffi[58])

With known coefficients α_i, it becomes possible to calculate the interaction energies $U(r_i)$ between two solute atoms which are i-th nearest neighbours by using an appropriate statistical model, such as that given by Clapp and Moss[111]. Based on this model, Chen and Balluffi[58] and Machlin[110] calculated the pair interaction energies in Pt-0.63 at% Ni and in Pt-3.4 at% Ni, respectively; both results agree quite well (Fig. 4.2).

As outlined in Chap. 3.5 it is not possible to determine short-range order with either the atom probe FIM or the imaging atom probe.

4.3 Radiation Induced Defects

4.3.1 Some General Remarks

Many radiation induced defects, e.g. Frenkel pairs composed of a vacancy and a self-interstitial atom (SIA) can not be imaged in the transmission electron microscope

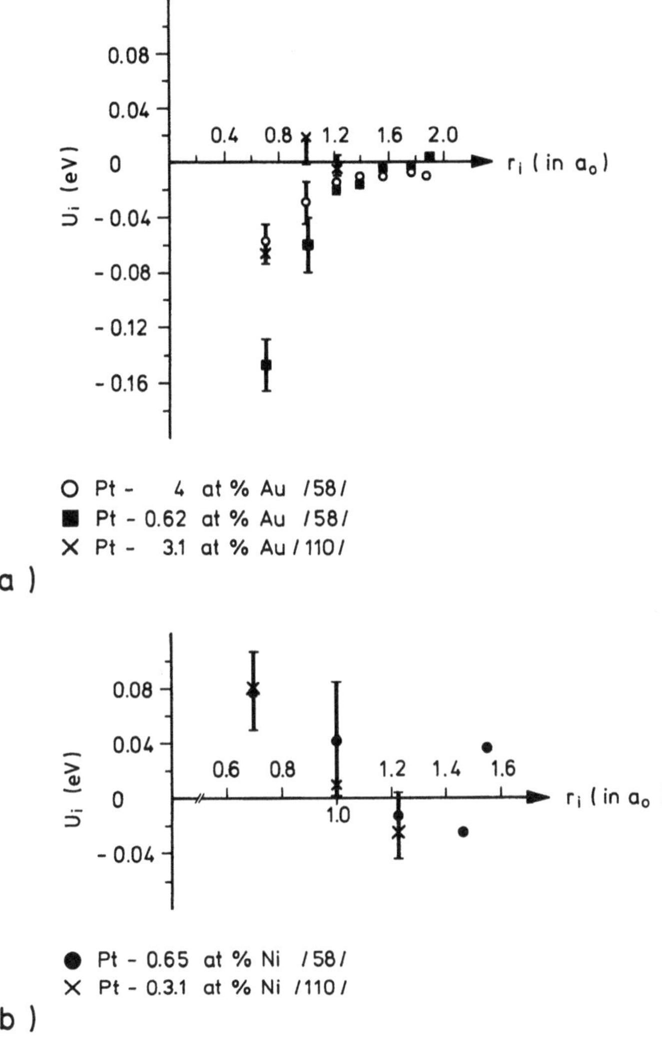

Fig. 4.2 a, b. Interaction energies between two gold atoms at different distances r_i in a platinum matrix calculated from the Clapp-Moss model (a); the same for two nickel atoms in platinum (b). (After C. G. Chen and R. W. Balluffi[58] and E. S. Machlin[110])

(TEM). Three-dimensional defects, such as depleted zones whose associated strain-fields are too small to produce sufficient strain contrast, also remain invisible in the TEM. This fact (and, of course, the technological importance of radiation damage) provided the impetus to apply field-ion microscopy very extensively and also very successfully to the study of radiation damage in metals. It is not possible to review in this chapter even part of the work which has been done in this area since 1960 and the interested reader is referred to review articles by Galligan[112], Müller[113] and especially to the more recent articles by Seidman and coworkers[54, 114–116], who have contributed most to the present knowledge of the spatial arrangement of point defects in irradiated metals.

The study of the defect structure in irradiated materials is essentially based on the atomic resolution capability of the FIM. Therefore, in connection with the discussion of Chap. 2 it is immediately obvious that FIM investigations of radiation damage are confined to either pure metals or a few dilute alloys. This fact, and the lesser susceptibility to the production of artifact vacancies of refractory metals, explains why numerous FIM studies of radiation defects have focussed on pure tungsten or dilute tungsten based alloys whereas almost no studies have been reported dealing with materials, e.g. ferrous alloys, which are more relevant for technological applications in environments exposed to irradiation.

In the following sections, we shall discuss a few recent applications and some results, which are typical of FIM studies of radiation damage and which have contributed much to the understanding of the defect structure in irradiated materials.

4.3.2 Defect Structure Generated by Ion Impact

Analyses of the defect structure in many metals (e.g. W, Pt, Ir) irradiated with various ions, including W^+, Mo^+, Cr^+-ions of energies in the range 20 to 30 KeV[55, 114–117], noble gas ions (Ar^+, Xe^+, He^+) with energies of 80 to 300 KeV[118, 119] and electrons of 1 to 2.5 MeV[118, 120] have been performed up to now. The chosen ion doses have ranged between 1×10^{11} and 5×10^{13} ions/cm^2. In most cases the irradiation has been performed *in-situ* in the FIM always in the absence of an applied field at a well defined tip temperature T_{irr}. The particular choice of T_{irr} is governed by the damage structure in which one is interested. If the analysis of the primary damage state prior to any recovery, i.e. recombination of Frenkel pairs, is the objective of the investigation, then T_{irr} has to be lower than the onset temperature (T_I) of stage I of the recovery spectrum; for pure tungsten T_I has been determined to be 28 K[121]; hence, $T_{irr} = 15$ K as chosen frequently by Seidman and co-workers[121, 122] is sufficiently low to maintain the spatial arrangement of the vacancies and SIAs in the lattice as generated by the primary knock-on atom (PKA). In contrast, during irradiation of tungsten at 80 K (i.e. in stage II according to Ref. 122), as recently performed by Stiller and Norden[119], the observed defect structure already reflects some recovery and clustering of SIAs due to the mobility of the latter above ~ 28 K[121].

To analyse the detailed defect structure after irradiation, the field ion tip is dissected (at either T_{irr} or at even lower temperatures) atom-by-atom (see Fig. 2.16) by slow pulsed field evaporation and subsequently isometrically reconstructed in three dimensions[123]. Such an analysis sometimes necessitates the recording of as many as 5×10^5 FIM images on film[55] in order to obtain statistically meaningful results. It was therefore necessary for Seidman and coworkers to develop an automated data recording and analysing system[124, 125].

By means of FIM it has been established, that in tungsten bombarded with ions of low doses at temperatures below 473 K, i.e. at approximately the end of stage II, most vacancies are concentrated in three-dimensional depleted zones[116, 117, 119], each of which is created by a single incident energetic PKA[115]. Because of the small elastic strain associated with these vacancy-rich depleted zones, they can not be detected in the TEM until they collapse into dislocation loops with sufficient associated strain.

As has been pointed out by Seidman[116], it is the existence of the depleted zones which explains the puzzling effect, that less vacancy contrast effects in the TEM are observed than one would have expected from the number of incident ions per unit area.

Figure 4.3 shows an isometric reconstruction of the vacancy distribution within a depleted zone in tungsten, created by in-situ bombardment at 10 K with 30-KeV W^+ ions[117]. Each open circle represents one vacancy and the connecting bonds between two adjacent vacancies correspond to first nearest-neighbour distances. In this depleted zone, which was detected about 8 nm below the irradiated surface, about 83% of the total number of vacancies was found in clusters composed of more than six first-nearest neighbour vacancies, whereas only about 13% appeared as monovacancies having no other vacancy as a first-nearest neighbour; the average diameter and vacancy concentration of this depleted zone were determined to be about 1.8 nm and 21 at%, respectively. Seidman[115], and recently also Stiller and Nordén[119], have shown that the morphology of depleted zones is basically independent of the chosen irradiation temperature ($T_{irr} \lesssim$ 473 K) and that in most cases they are extended in $\langle 110 \rangle$-directions, giving some direct evidence for channelling events. After low temperature irradiation ($T_{irr} \lesssim$ 18 K) of tungsten, Seidman[54, 115] also reports the observation of some SIAs at a distance of 4.5 to 15 nm and 4.5 to 8.5 nm along $\langle 110 \rangle$ and $\langle 111 \rangle$, respectively, whereas after irradiation at 473 K no SIAs were detected[115]. It has been emphasized[54], that for geometrical sampling reasons the measured distances of the SIAs need not necessarily be lower and upper bounds for the range of focused replacement collision sequences (RCS). The results, however, show that the range of RCSs seems to be considerably greater than is evaluated by computer simulations[126].

In recent work, Wei and Seidman[117] have investigated for the first time the structural variations of depleted zones in tungsten created by 30 keV ions of different masses, e.g.

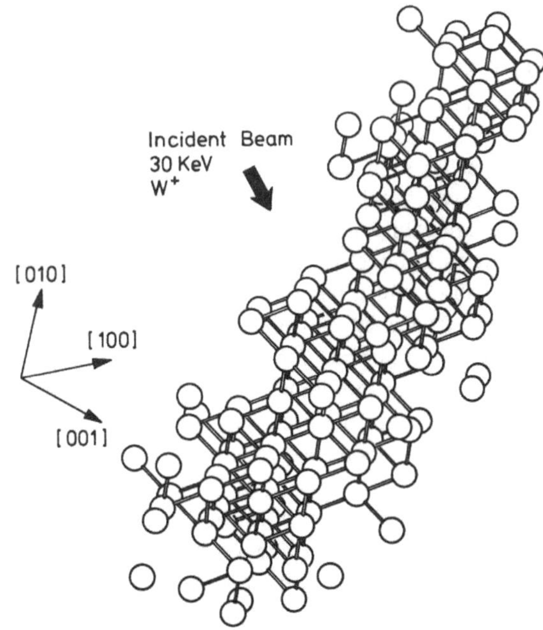

Fig. 4.3. Isometric reconstruction of the distribution of vacancies within a depleted zone in tungsten irradiated with 30 keV W^+ ions. (After C. Y. Wei and D. N. Seidman[117])

W^+ (~ 186 a.m.u.), Mo^+ (~ 96 a.m.u.) and Cr^+ (~ 52 a.m.u.). They showed, that with decreasing incident ion mass, the spatial extent of the depleted zones increases (e.g. 4.2 nm for Cr^+-irradiation) and that there exists a tendency for the depleted zones, which are rather compact if generated by W^+ (Fig. 4.3), to form subcascades. Furthermore, the vacancy concentration and the number of larger clusters composed of first-nearest neighbour vacancies is also decreased, if the depleted zones are created by projectiles of lower mass.

In a similar type of analysis, the same authors[123] detected a platelet-shaped depleted zone with a (220) habit plane in Pt-4 at% Au irradiated with 30 keV W^+ ions at 40 K, isochronally warmed to 100 K and subsequently imaged and dissected atom-by-atom at 40 K. The disc-shaped depleted zone, which contained approximately 40 at% of vacancies, was concluded to represent an intermediate energy state between the higher-energy vacancy configuration of a depleted zone and that of a lower-energy prismatic dislocation loop. The latter, which are formed by a collapse of depleted zones, can sometimes be encountered in the TEM but are lost for FIM examination since they can glide out of the near-by tip surface under the applied field (Chap. 4.1) as has been verified in a recent combined FIM and TEM analysis of tungsten irradiated with 180 keV Xe^+ ions[119]. A TEM examination of the irradiated tip, prior to FIM imaging is therefore strongly recommended in cases where warming the bombarded tip to room temperature can be tolerated.

It should be pointed out that, apart from computer simulations, such a comprehensive and detailed insight into the defect structure of irradiated metals could, and still can, only be gained from field-ion microscopy. Thus, the FIM provides the only means for checking to what extent the physical models entering the computer simulations of the primary state of damage are realistic.

4.3.3 Radiation Damage Created by Neutrons

The FIM technique has also been applied frequently to the study of defects in both pure metals (e.g. Pt[127, 128], Ir, Mo[129], W[118, 130]) and alloys (e.g. Fe-0.34 at% Cu[81] or Mo-1 at% Ti[131]) after fast neutron irradiation or after irradiation of tungsten with thermal neutrons (E < 50 eV). In general, all n-irradiations have been performed at temperatures between 60 and 700 °C, i.e. at temperatures where long-range migration and, hence, annealing of SIA already occurs. Therefore, most investigations have focussed on the analysis of agglomerates of vacancies, such as depleted zones, voids and dislocation loops, all of which have been shown by FIM to exist in n-irradiated materials. In most studies of vacancy clusters in n-irradiated materials, the dissection increment during pulsed field evaporation of the tip has been too large to obtain a detailed picture of the arrangement of individual vacancies within the various vacancy agglomerates, as has been described in the previous section for ion-irradiation. Therefore, the analyses have been essentially confined to the determination of the size distribution of the particular defects.

Figure 2.19 shows part of the dissection of a Fe-0.34 at% Cu tip irradiated with fast neutrons at 288 °C to a fluence of 3×10^{19} n/cm^2 [81]; the chosen dissectional increment was one {110} plane, corresponding to a depth probing of 0.2 nm between each frame (Chap. 2.7.1). In Fig. 2.19 three microvoids are seen having diameters of 0.6 ± 0.2 nm.

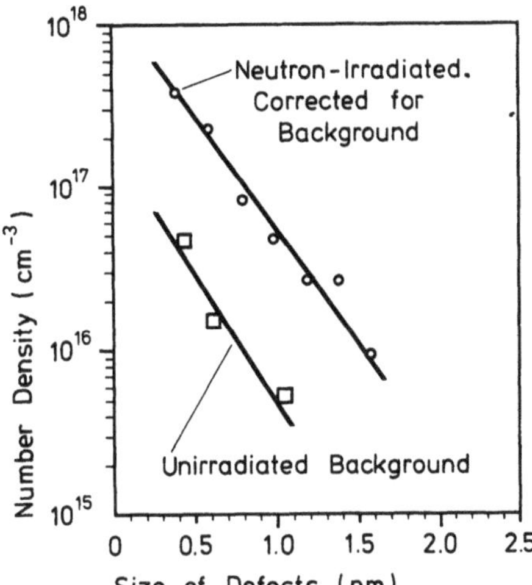

Fig. 4.4. Size distribution of microvoids in neutron-irradiated ($3 \times 10^{19}\,\text{n/cm}^2$; > 0.1 MeV) Fe-0.34 at% Cu[81]

As is depicted in Fig. 4.4, the number density of the voids is about $8 \times 10^{17}\ \text{cm}^{-3}$ with diameters ranging between 0.4 and 1.6 nm; such small voids can not be detected in the TEM. It was proposed[81] that the observed microvoids in n-irradiated Fe-0.34 at% Cu are stabilized against dissolution or collapsing into dislocation loops by an energetically favourable copper segregation to the void surface. However, in a recent quantitative investigation of solute segregation to the void surface by means of the atom probe, Wagner and Seidman[131] did not observe any significant segregation of Ti atoms to voids in a Mo-1 at% Ti − 0.24 at% C alloy after n-irradiation ($\sim 1 \times 10^{22}$ n/cm^2, E > 1 MeV) at 700 °C, but rather found some evidence for the segregation of interstitial carbon. The observed carbon segregation is thought to enhance the void nucleation rate and, hence, to increase the volume density of the voids[131].

Although these FIM and atom-probe studies have to be considered as preliminary, they have confirmed that these techniques can contribute significantly in the future to clarify the effects of solute and interstitial segregation on the formation and/or stabilization of voids and their influence on the swelling susceptibility of alloys under irradiation; further, one might expect that the imaging atom probe will turn out to be even more powerful than the atom probe for such investigations.

4.3.4 Analyses of Recovery after Irradiation

The FIM has frequently been used to analyse the recovery spectra of irradiated metals and alloys. Much emphasis has been placed on the question of whether or not the motion of a single defect type is responsible for each of the well-defined stages of the recovery spectrum as usually revealed by the resistivity changes occuring during isochronal warming experiments. In addition, much interest has been focussed on the role that impurities play in the recovery process. Of particular concern is the mechanism of recovery in Stage III; this interest results from the long-term controversy over whether the kinetics of Stage III are controlled by the long-range migration of vacancies, or by the migration of a second type of self-interstitial, as postulated in the 2-interstitial model by Seeger[132, 133]. The Stage-III SIA is proposed to be a thermally-converted Stage-I SIA of lower energy with a substantially reduced rate of migration.

Analyses of the recovery behaviour in Stage I, II and III have focussed on a variety of metals and alloys, including undoped tungsten[54, 55, 121, 130] and platinum[134], Pt-Au alloys[134, 136] as well as tungsten doped with some carbon[122] or rhenium[122, 135]. These materials have been irradiated with either ions, neutrons or electrons[120]. Depending on the particular recovery stage under investigation, the irradiation temperature has been chosen to be sufficiently below the onset temperature of this stage, which usually is known from the recovery of the resistivity; following irradiation, the FIM specimens are warmed at a constant rate. Since the tip surface acts as an excellent sink for SIAs, the flux of SIAs to the specimen surface can be directly observed over a given temperature interval and both the enthalpy of migration and the diffusivity of a SIA can be evaluated[121, 134].

Figure 4.5 shows four isochronal annealing spectra obtained between 18 K and 120 K by Wilson et al.[122] for pure tungsten, tungsten doped with carbon and for two different W-Re alloys, which had all been irradiated at 18 K with 30 keV W^+ ions to a dose of 5×10^{12} ions/cm^2. The spectra are presented as the fraction of the total number of SIAs which appeared as bright spots on the tip surface (Chap. 2.6.4) in the given temperature interval. It was found that the recovery spectra for tungsten *do not* depend significantly on its impurity level. Basically, they exhibit one dominant peak at 38 K due to long-range migration of Stage I SIAs and some more or less well defined recovery peaks at around 50 K, 60 K, 80 K, 95 K and 110 K[54, 122]. These structural features of the recovery spectrum are essentially maintained after doping with $\sim 10^{-2}$ at% C (Fig. 4.5 b), whereas an addition of 0.5 at% Re shifts the main peak, associated with the long-range migration of Stage-I SIAs, to about 33 K (Fig. 4.5 c); this shift has been explained to be caused by SIAs lying sufficiently close to the tip surface to avoid trapping by Re atoms during their migration to the surface. In pure specimens those SIAs within close proximity to the surface mainly govern the SIA flux observed below the peak temperature of 38 K (Fig. 4.5 a). Wilson et al.[122] attributed the dominant peak at 50 K in the W-0.5 at% Re alloy to Substage II a of the recovery spectrum. In contrast to this alloy, the more concentrated W-3 at% Re alloy yielded almost no SIA recovery between 18 and 120 K (Fig. 4.5 d), suggesting that most SIAs have been trapped at Re atoms already in Stage I.

From these experiments, the conclusion was drawn, that during the long-range migration in Stage I, most individual SIAs are trapped in SIA clusters rather than at impurity atoms, and only at extremely high impurity levels, e.g. 3 at% Re, does the formation of strongly bound *SIA-impurity atom agglomerates* become the dominant trapping mechanism. Therefore, the early Stage-II SIA recovery observed in undoped tungsten of various impurity levels as well as in the dilute W-C and W-0.5 at% Re alloy has been associated with the migration and/or dissolution of SIA clusters which have been formed during Stage I. However, since in the concentrated W-3 at% Re alloy most SIAs become strongly bound to Re atoms, only few SIA clusters can be formed in Stage I and, hence, almost no recovery is observed in the temperature range between 45 and 120 K.

Similar experiments, which have also yielded direct evidence for uncorrelated long-range migration of SIAs in Stage I and early Stage II, have been performed in Pt and various Pt-Au-alloys[134, 136]. The recovery behaviour in early Stage II of Pt-Au, i.e. between 40 and 100 K, shows a delayed diffusion of the SIAs due to a continuous

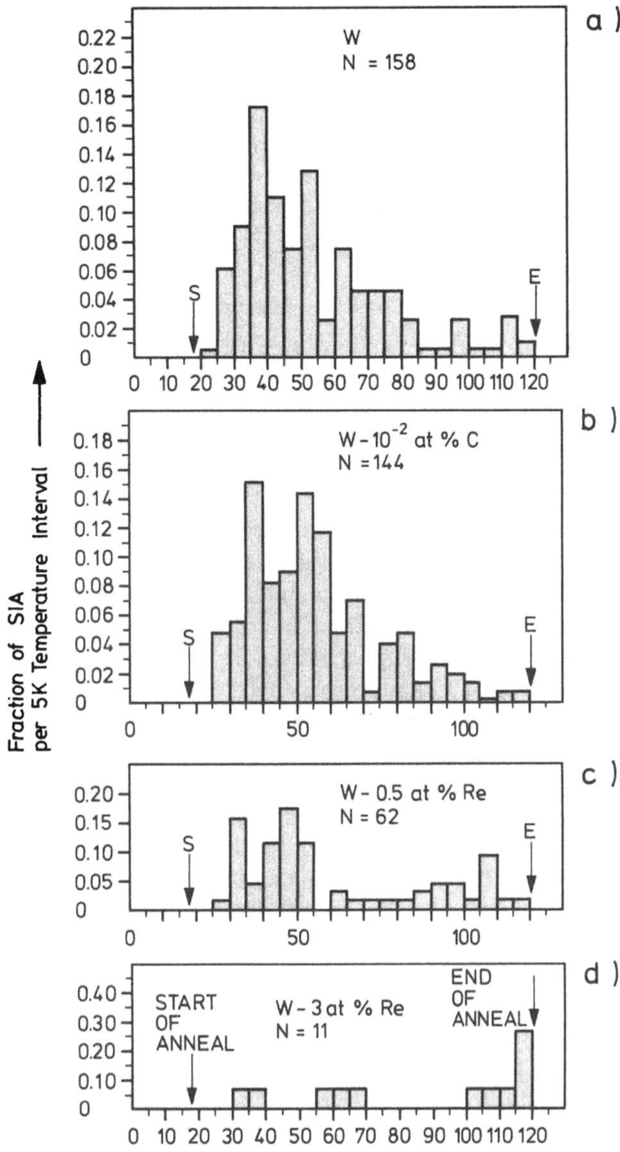

Fig. 4.5a–d. Composite isochronal annealing spectra for tungsten with different impurity levels. *N* indicates the total number of detected defects. All specimens were irradiated at 18 K with 30 keV W^+ ions to a dose of 5×10^{12} ions/cm^2 and subsequently annealed to 120 K at a warming rate of about 2.5 K/min. (After K. L. Wilson, M. I. Bakes and D. N. Seidman[122])

trapping and detrapping of the SIAs at and from Au atoms until the SIAs have finally reached the surface.

As has been pointed out above, the controversy regarding, whether Stage III (above 500 K) in the recovery spectrum of tungsten is to be attributed to the annealing of a thermally converted "Stage-III SIA" (a split SIA) or to the migration of monovacancies, has initiated several FIM investigations which have attempted to clarify this point by direct observation of the defect flux. Early FIM investigations by Attardo et al.[137, 138] reported the observation and, hence, the existence of a split SIA in Stage III of neutron irradiated tungsten, and also the occurrence of a Stage IV which is thought to be caused by the annealing of vacancies. In a recent investigation of the isochronal annealing of tungsten of various impurity levels irradiated with *thermal* neutrons, Kim and Galligan[135] basically reconfirmed their earlier observations and conclusions that Stage III recovery is dominated by SIA migration with an activation energy of about 1.71 eV; however, Stage IV was only observed in their purest tungsten specimen, containing an impurity level of about 1 at. ppm, and was attributed to the dissolution of SIA clusters formed during Stage III.

In contrast to these FIM experiments, Seidman and coworkers[54, 116, 120, 139] did not find any evidence for isolated Stage-III SIAs in tungsten irradiated with electrons at 430 K, but rather found apart from immobile monovacancies another immobile defect exhibiting a complex contrast pattern and extending over several atomic layers. This contrast pattern was interpreted to be due to a cluster composed of SIAs *and* impurity atoms. This conclusion was supported by the observation of similar contrast patterns in the above mentioned W-3 at% Re[114, 122] ant Pt-Au alloys[136], which could be identified as SIA-Re complexes and SIAs trapped at individual Au atoms, respectively. Based on his own experiments and given support from resistivity experiments, Seidman[54, 139] critically commented the experiments and conclusions drawn by Attardo et al. and Kim and Galligan. He pointed out that in order to identify an individual SIA unequivocally from the observed FIM contrast pattern, it is necessary to dissect a high index plane atom-by-atom (Chap. 4.3.2), a process which often necessitates the examination and analysis of some hundreds of film frames for one single contrast pattern. Thus, the three-frame dissection sequence presented by Kim and Galligan[135] is not sufficient to attribute the observed contrast pattern unequivocally to a Stage-III SIA, since the contrast could also have arisen from a SIA trapped at a Re atom already during Stage I; the Re traps have been "alloyed" to the tungsten specimen due to transmutations during thermal neutron irradiation. The interpretation that the observed contrast effect was caused by a SIA-Re complex rather than by a split SIA was further supported by the fact that the activation enthalpy of migration of 1.71 eV, as measured by Kim and Galligan for Stage III recovery, is identical to that for the migration of a monovacancy[139]. Therefore, it must be concluded that Stage III is actually dominated by the migration of monovacancies and that to date none of the FIM analyses of Stage-III recovery have yielded any convincing evidence for the existence of a Stage-III SIA.

Doubts are sometimes cast upon the significance of the iso-chronal recovery spectra as determined by FIM and, hence, especially on the determination of the diffusivity D_{SIA} of a Stage-I SIA from those spectra, since D_{SIA} is influenced by the applied electric field F as we have discussed in Chap. 2.5. This influence of F on D_{SIA} has been examined by Seidman and coworkers quite comprehensively[121, 140]. They re-evaluated recently the volume change of migration ΔV^m of the stage-I SIA in tungsten to be less than 0.02 atomic volume[140] (as compared to a former estimate of $\Delta V^m \leq 0.1$ atomic volume[121]). Then, according to Eq. (2.6b), the internal energy of migration ΔU^m of the SIA is reduced by less than 0.02 eV if the applied field is assumed to be 47.5 V/nm during imaging; by comparison the enthalpy of migration of the SIA in tungsten, which is the measurable quantity under an applied field, was determined to be $\Delta H^m \approx 0.085$ eV[123]. Therefore, the effect of the applied field is to shift the onset of SIA migration, i.e. Stage I, towards lower temperatures as compared to zero-field migration; as was pointed out recently by Seidman[116], there is some experimental evidence that the temperature shift of the main peak in the Stage-I recovery spectra, caused by the electric field, ranges only between 5 and 6.5 K. Thus, SIA migration only starts during the warming experiments und not at temperatures below 18 K where most irradiation experiments have been performed for the analyses of Stage-I and Stage-II recovery.

4.3.5 Composition Profiles of Implanted Species

Only recently both the atom-probe FIM and the imaging atom-probe technique have been applied to the study of the composition profile of deuterium[141] and helium[116] implanted into tungsten. The implantations of deuterium and helium ions into the tungsten tips were performed by *in-situ* irradiation at 300 K and 80 K, respectively. The chosen energies (80 eV for D^+ and 300 eV for He^+) of the implanted ions are not sufficient to create point defects in tungsten and, hence, the only defects after irradiation are interstitial deuterium and helium atoms. From the composition profiles as determined by the atom-probe FIM to a depth of about 16 nm, Wagner and Seidman[116] evaluated the mean value of the He range profile (defined as the first derivative of the S-shaped He composition profile) to be about 4.7 nm and the standard deviation to be 3.5 nm. Furthermore, they showed that interstitial He becomes mobile between 80 and 110 K. The presumable high mobility of interstitial deuterium below 300 K might be one explanation for the comparatively high concentration of D in the top-most surface layer as determined by Panitz[141] by means of the imaging atom probe, although the possibility that the deuterium ion beam was not monoenergetic, leading to the retention of some extremely slow ions at the surface can not be entirely ruled out.

These studies must still be considered as preliminary; however, they have shown that the atom-probe depth profiling technique may be expected to be quite powerful in future analyses of the range profile of gas atoms, which are injected into various materials in order to simulate the behaviour of these materials under a thermonuclear environment.

4.4 Disorder-Order Transformations

As has been pointed out in Chap. 2.6.2, the degree of regularity in the FIM image of an alloy undergoing a disorder-order reaction is a qualitative measure of the degree of order. Consequently, the FIM has been used in the past quite frequently to follow disorder/order reactions in various alloys including Ni_4Mo[142–148], Ni_4W[147], Pt_3Co[149], Ni_3Ti[150, 151], Ni_3Fe[152], Ni_2Cr[153], Ni_2V[153] and Pt-Co[149, 154]. The emphasis of these FIM investigations has chiefly concerned the following questions: i) does ordering, i.e. the formation of the superlattice, occur via a thermodynamically first order reaction with the formation of nuclei of the completely ordered phase which subsequently grow into the disordered matrix until they meet other growing ordered domains, or ii) is the ordering transformation a continuous reaction, in which the degree of order within each domain changes with time (as a second order reaction)? iii) Which types of domain boundaries do occur?

The most systematic FIM investigations have been performed in recent years by Yamamoto and coworkers[144–148] mainly on a stoichiometric Ni_4Mo-alloy. Figure 4.6 shows this alloy in the disordered state after quenching from 1100 °C. There are two characteristic features observed: Firstly, only low index poles are relatively well developed, and secondly, there exist numerous bright clusters of about 1 to 3 nm in diameter, which are randomly distributed. Computer-simulated images of *ideally disordered* Ni_4Mo, making the assumption that only Mo atoms are imaged (see Chap. 2.6.2)[148], reveal clearly the regions of low index poles, but they exhibit much fewer bright-spot clusters than observed in Fig. 4.6. This discrepancy between the contrast pattern of the simulated, and the actual FIM image has been attributed to the existence of small short-range ordered Mo-rich regions having diameters between 1 and 3 nm; the existence of such short-range ordered regions in the disordered alloy has also been concluded from TEM investigations[155].

Figure 4.7a depicts the same alloy after subsequently heating *at a high rate* to 800 °C and annealing for 5 min. Several large long-range ordered domains can be seen embedded in the disordered α matrix. In this case, ordering takes place via the nucleation and growth (i)) of ordered domains. However, by choosing a much *lower heating rate* with otherwise unchanged annealing conditions the statistically *homogeneous ordering reaction* (ii)) becomes operative resulting in many small ordered domains (Fig. 4.8),

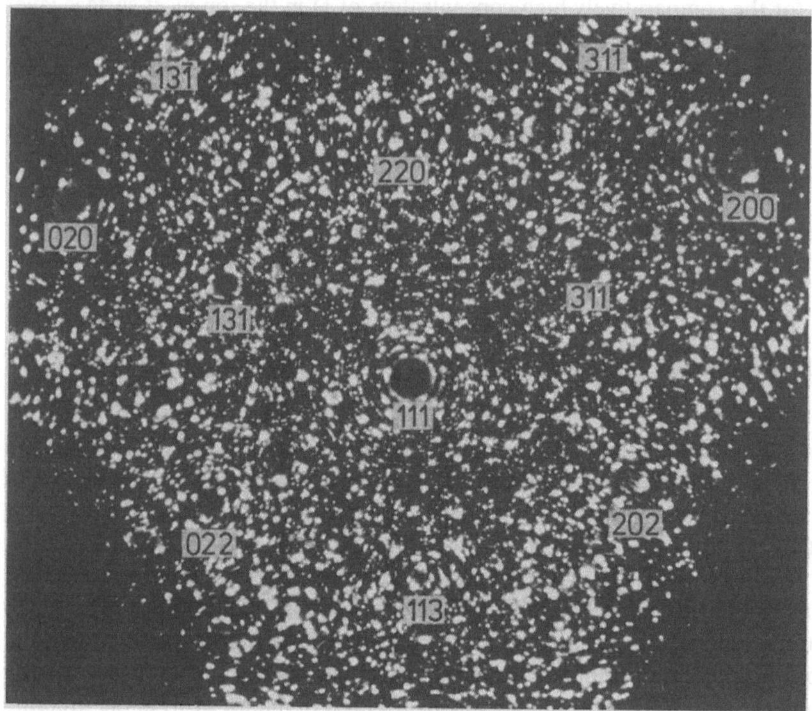

Fig. 4.6. Field-ion image from an as-quenched Ni-20 at% Mo alloy exhibiting some short-range ordering (clusters of bright spots of diameters ranging between 1 and 3 nm); the low index poles are still to be recognized. Imaged in helium at 21 K. (Courtesy M. Yamamoto)

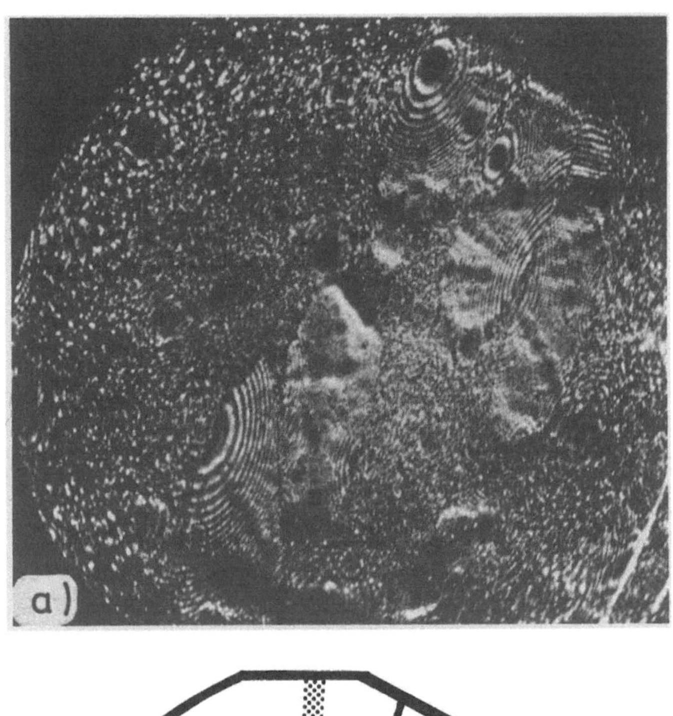

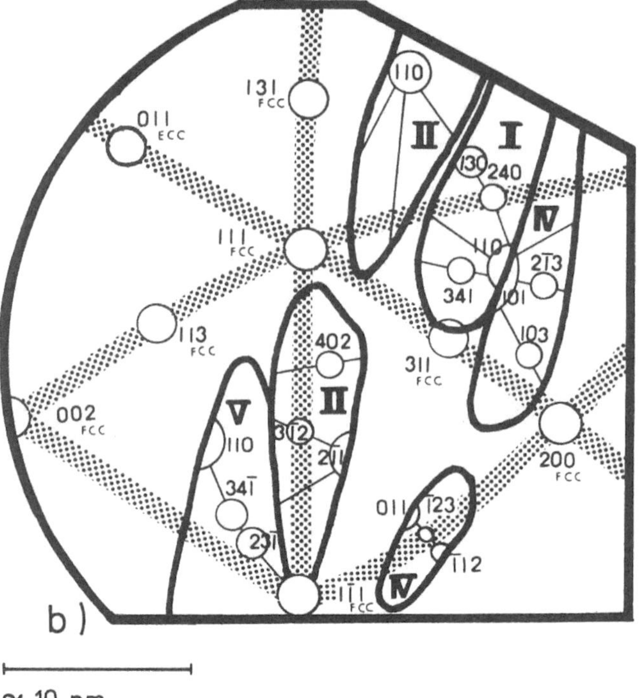

Fig. 4.7a, b. The same alloy as in Fig. 4.6 after annealing for 5 min at 800 °C. Some ordered domains are embedded into the disordered matrix (**a**). In (**b**) the relative size of ring steps and the $\langle 110 \rangle_{fcc}$ dark bands (*dotted*) are shown schematically. The ordered domains are labeled according to Table 4.2. (Courtesy M. Yamamoto)

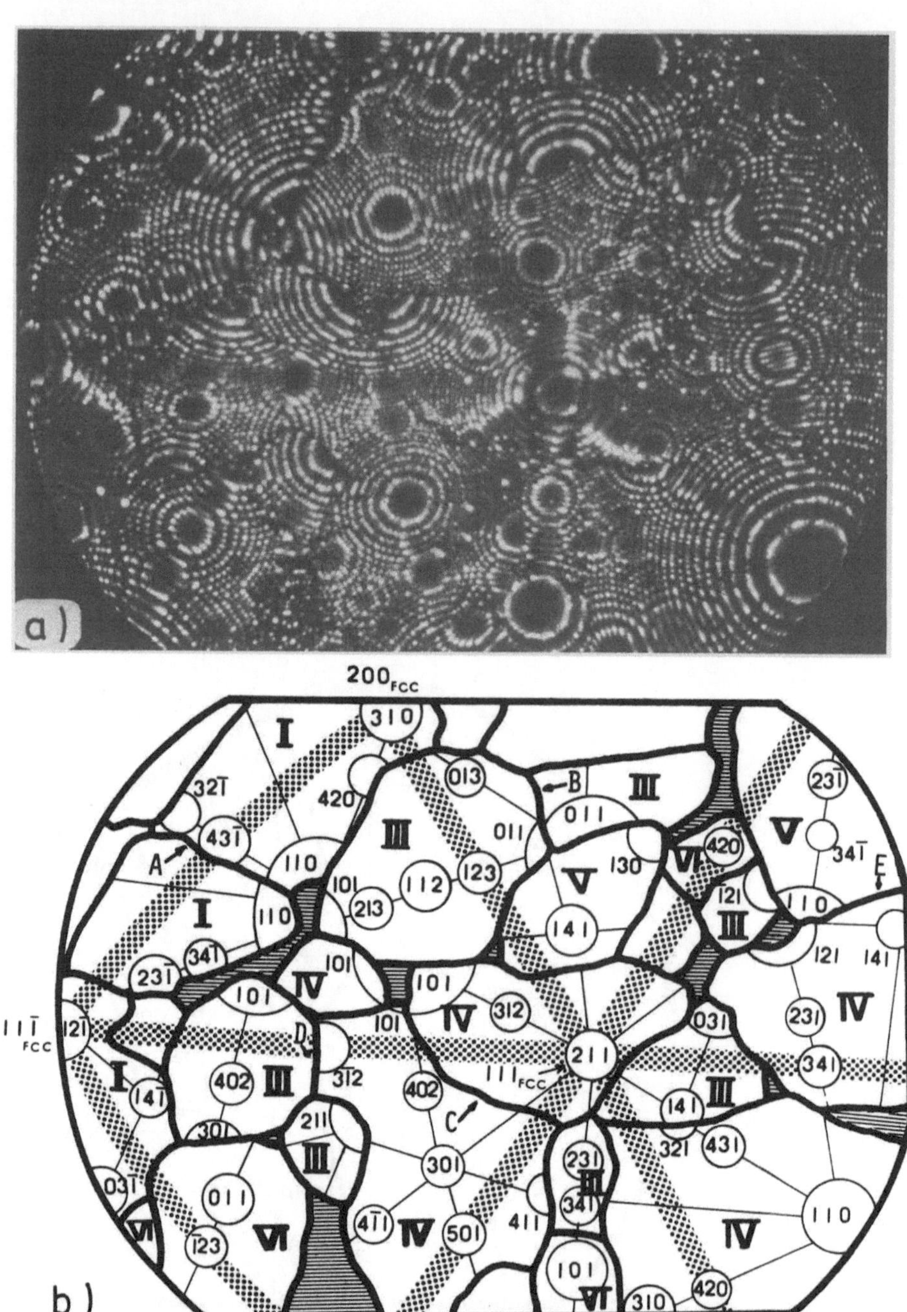

Fig. 4.8 a, b. The same alloy as in Fig. 4.6 after annealing for 30 min at 800 °C. The recognizable domain structure is schematically represented in **b)** together with $\langle 110 \rangle_{\text{fcc}}$ dark bands. The hatched regions correspond to the disordered α phase; poles without a suffix "fcc" are indexed with respect to the bct lattice of the ordered β phase. (Courtesy M. Yamamoto)

homogeneously distributed throughout the specimen and separated by domain boundaries. Computer-simulated FIM images of Ni_4Mo undergoing a homogeneous ordering transformation show clearly[148] that the principal superlattice poles of the ordered bct β phase, e.g. $(011)_{bct}$, develop gradually with the degree of order; hence, for instance by observing the development of the $(011)_{bct}$ pole, it becomes possible to distinguish a second order ordering reaction from a first order one.

Although not yet performed, it should also be possible to determine the degree of order in all stages of a continuous transformation by comparing the simulated images of given ordering parameters with those of the actual field-ion images.

Similar FIM studies have been performed in Ni_3Fe and Ni_2Cr annealed at 480 and 500 °C, respectively, by Taunt and Ralph[152, 153]. From the absence of any two-phase structure during the entire course of the ordering reaction, it was concluded that at these annealing temperatures ordering occurs in both alloys via a continuous (or homogenous) reaction rather than by classical nucleation and growth. These authors tried to follow more quantitatively the development of modulations in the degree of local order within the short-range ordered state in Ni_3Fe[152]. For this purpose, they employed the so-called *plane stability ratio technique*[150], which is based on the fact that the relative stability to field evaporation of successive superlattice planes, e.g. the {200} planes in Ni_3Fe, depends on the chemical composition of that plane, i.e. on the degree of order[143]. In practice, a constant voltage, which is just above that necessary for field evaporation, is applied to the tip, and the times of collapse of the successive {200} superlattice planes are recorded as a function of depth x_{200} along $\langle 200 \rangle$. The plane stability ratio, which is a *qualitative* measure of the local degree of order, is defined as[150, 152]

$$R = \log \frac{t_n}{t_{n+t}} \tag{4.3}$$

with t_n being the time of collapse of the n-th {200} superlattice plane; in a completely disordered phase, R is equal to 1 whereas in ordered regions it is defined to be < 1. In principle, this technique is analogous to the determination of the concentration profile in an alloy using the atom probe (see Chap. 3.2) with the plane stability ratio R being replaced by the planar concentration. Hence, to evaluate any order modulations in the R(x) profile the same types of analysis, i.e. autocorrelation and Fourier analysis, as described in Chap. 3.2 for the concentration profiles c(x), can be applied to R(x). Having performed these analyses, Taunt and Ralph[150] found in Ni_3Fe already after the quench a dominant wavelength of the order modulations of about 1.5 nm, thus confirming the interpretation of Yamamoto for Ni_4Mo[147] that there exist already some deviations from randomness in the as-quenched specimens. After a subsequent anneal at 480 °C the dominant modulation wavelength in Ni_3Fe increased linearly with aging time to (8 ± 3.5) nm; at this transformation stage, the modulation wavelength is identical with the domain size.

It is known[146, 155], that in Ni_4Mo only six out of the thirty possible ways, by which the nucleation of the ordered β phase (bct) in the α matrix (fcc) can occur, result in a different orientation of the superlattice with respect to that of the α phase. The domain boundaries which are formed after two ordered domains of different orientation have impinged, can be either of the translational antiphase boundary type (APB) or different types of twins, i.e. antiparallel twin boundaries (APTB) or perpendicular twin boundaries (PTB)[155]. Once the type of two adjacent domains is determined by their crystallographic relationship with respect to α, the type of domain boundary separating them can also be determined.

In principle, the type of domain and domain boundary can be determined by means of TEM and electron diffraction since each of the three different types of domain boundary (APB, APTB and PTB) gives a specific fringe contrast pattern[155]. However, it becomes very difficult, if not impossible, to identify the various domains and domain boundaries in the TEM if the domains are small, i.e. in the order of only a few nm. In this case the contrast from an individual domain is obscured by the surrounding domain structure; this holds especially true if there is a strain field associated with the domain boundary such as for a PTB boundary in Ni_4Mo[155]. This problem can be circumvented by using FIM as is shown in Fig. 4.8, where several impinging domains can be seen. In order to identify the boundaries between adjacent domains, it is first necessary to determine the crystallographic relationship between the two domains.

Early attempts to analyse domain boundaries in Ni_4Mo using FIM[156, 157] only allowed translational APBs to be unequivocally identified. These analyses were essentially based on the application of the $g \cdot R$ criterion[158] also used in the TEM (where g is a known reciprocal lattice vector and R the displacement vector of the antiphase boundary), or, on the evaluation of the orientation changes and the ring mismatching across the interface. Only in combination with TEM were Lefevre et al.[156, 157] and Chakravarti et al.[159] able to identify APTB and PTB in the FIM patterns of ordered Ni_4Mo. A method for distinguishing between the three types of interface, which is independent of complementary electron microscopy and which considers the matching of prominent planes across the boundaries has been proposed by Chandrasekharaiah and Ranganathan[142]; however, until now this theory has not yet been shown to provide easy access to the identification of the different domain boundaries. An alternative method for the identification of domains and, hence, the type of interfaces in the FIM patterns of Ni_4Mo was elaborated recently by Yamamoto et al.[145]. This method is based essentially on the occurrence of the following peculiar features inherent to field-ion images of Ni_4Mo[145], some of which are easily recognizable in Fig. 4.7 and 4.8: i) The FIM images of individual ordered domains clearly reveal bct symmetry. ii) As was already noticed by Newman and Lefevre[157, 160], the regions around both the $\{111\}_{fcc}$ (Fig. 4.7 and 4.8) and $\langle 110 \rangle_{fcc}$ zones (dotted bands in Fig. 4.7b and 4.8b) appear consistently darker than other regions through all stages of ordering; these dark $\{111\}_{fcc}$ regions, which are presumably caused by an increased local radius of curvature, correspond to dark $\{121\}_{bct}$ regions in completely ordered specimens. iii) The relative size of ring steps of various poles $\{hkl\}_{bct}$ decreases in the order of decreasing interplanar spacing d(hkl). This is quite evident in Fig. 4.8a and is considered schematically in Fig. 4.8b as circles of three different diameters. iv) Yamamoto et al. were also able to distinguish between two poles in the bct lattice by considering the arrangement of bright spots in the particular $(hkl)_{bct}$ plane. For example, it was found that in $\{411\}_{bct}$ planes each bright spot is well separated from the adjacent one whereas in $\{310\}_{bct}$ planes the bright spots overlap and form bright rows rather than single bright spots; however, this feature is not as evident in Fig. 4.8a as the previous ones. v) By virtue of Eq. (2.8) and by following the procedure outlined in Chap. 2.7.1, i.e. by counting the number n of rings of net planes, the local radius of curvature r_t can be determined, provided that two poles $(hkl)_{bct}$ and $(h'k'l')_{bct}$ are identified, and that d_{hkl} is replaced by the appropriate interplanar spacing $S_{hkl} \geq d_{hkl}$ of only the visible planes. (This substitution becomes necessary since only Mo atoms are imaged and, hence, e.g. superlattice planes composed of Ni atoms do not reveal a ring in the FIM pattern as has been discussed in Chap. 2.6.2.) If the plane (h'k'l') is determined correctly, then r_t ought to be approximately equal to the average tip radius $r_t = V_0/\varkappa \cdot F_{BIV}$ (see Chap. 2.7.2) with $\varkappa = 4.9$ as determined by Yamamoto et al.[146].

In order to identify individual domains in the FIM pattern, six different charts for the six possible domain orientations are constructed (e.g. Fig. 4.9). (Actually, by taking into account the crystallographic relations between the six different domains, three of them, at a time, can be brought into coincidence by an appropriate rotation around the $(111)_{fcc}$ pole; hence, only two charts are necessary[145].) These charts are essentially stereographic projections of the ordered bct lattice, but they also contain some poles indexed on the basis of the disordered fcc lattice (Fig. 4.9). Furthermore, these charts also depict the $\langle 110 \rangle_{fcc}$ dark lines (feature ii)) and the relative size of ring steps of various poles (feature iii)). By searching for coincidence between one of these charts and one particular domain of the FIM image, and, additionally, by taking into account the features i), iv) and v) it becomes possible to identify all domains observed in the field-ion image. As a

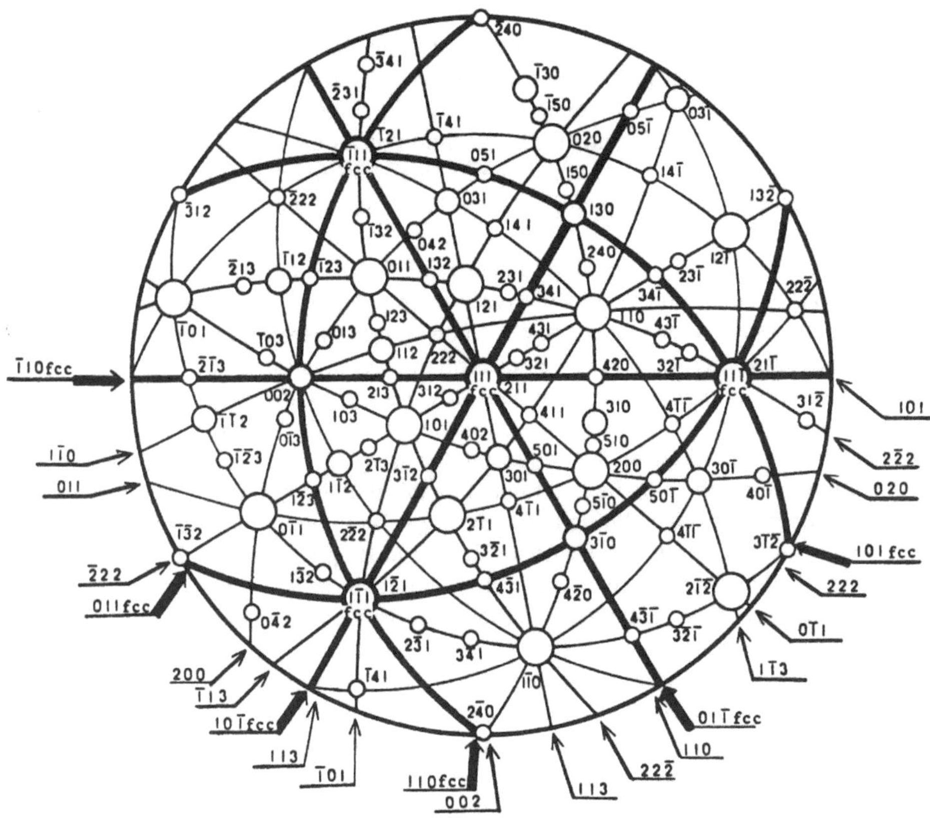

Fig. 4.9. Chart for domains II, IV and VI based on the stereographic projection of ordered Ni₄Mo with the $(111)_{fcc}$ pole at the centre. The poles without suffix "fcc" are indexed with respect to the bct lattice of domain II (After Ref. 145)

67

Table 4.2

Domain	I	II	III	IV	V	VI
I	APB[a]	APTB	PTB	PTB	PTB	PTB
II	APTB	APB[a]	PTB	PTB	PTB	PTB
III	PTB	PTM	APB[a]	APTB	PTB	PTB
IV	PTB	PTM	APTB	APB[a]	PTB	PTB
V	PTB	PTB	PTB	PTB	APB[a]	APTB
VI	PTB	PTB	PTB	PTB	APTB	APB[a]

[a] The probability that two impinging domains of the same type form a translational APB is 0.8, whereas the probability for meeting without a phase jump is 0.2. Furthermore, each APTB and PTB may be accompanied by a APB at a probability of 0.8[145]

result, Fig. 4.8 b shows that five different types of domains (I, III to VI) besides some disordered regions (hatched areas in Fig. 4.8 b) can be observed simultaneously in this Ni$_4$Mo-specimen which was annealed for 30 min at 800 °C. Once the domain structure is identified the type of domain boundary between two adjacent domains is immediately obtained by means of Table 4.2[145].

It is immediately evident from Fig. 4.8 b and Table 4.2 that all three types of domain boundary appear in equal proportions during the early stages of the ordering reaction. However, as was observed by Yamamoto et al.[147], coarsening of the domains during further annealing at 800 °C results in the retention of APBs and PTBs in preference to APTBs[147]. This result provides some experimental evidence that the surface energy of the anti-parallel twin boundary is higher than that of an APB or PTB as was pointed out by Ruedl et al.[155].

In contrast to TEM, it is possible to determine the local degree of order (S) and the type and size of small domains in the early stages of the ordering transformation by means of FIM, as the above results have demonstrated. Therefore, it might be expected that future applications of the FIM will focus on the investigation of the time evolution of S and the kinetics of domain growth in the early transformation stages, since little work has been done so far in this area. In particular, it should be possible to follow the ordering kinetics quantitatively by employing the atom probe; in principle, not only the modulation wavelength could be determined by this method but also the time evolution of the local degree of order during the early stages of a continuous ordering reaction.

In connection with ordering reactions, Taunt and Ralph[161] studied the fine structure of superdislocations of the $a\langle 110\rangle$ type in an intermetallic ordered Ni$_3$Al alloy. By means of an elaborated contrast analysis of the FIM patterns they showed the $a\langle 110\rangle$ superdislocation to be dissociated either into two $a/2\langle 110\rangle$ dislocations (i.e. two complete dislocations of the disordered phase) separated by a ribbon of antiphase boundary or into two $a/3\langle 211\rangle$ superlattice partial dislocations; however, both the $a/2\langle 110\rangle$ and the $a/3\langle 211\rangle$ superlattice partial dislocations did not show any evidence for a further dissociation either into two Shockley partials or into two dislocations with Burgers vectors of the type $a/6\langle 112\rangle$ and $a/2\langle 110\rangle$, respectively. From the measured dissociation width of the superlattice dislocation (about 3 nm), the antiphase boundary energy was evaluated to range between 0.25 and 0.35 Jm^{-2}. These values agree quite well with that of 0.3 Jm^{-2} as calculated from simple first nearest-neighbour bonding considerations[162].

These results demonstrate, that it is possible to study the fine structure of superlattice dislocations in ordered alloys in much detail by means of FIM; however, as was already pointed out in Chap. 4.1, the same type of information can often be obtained more easily in the TEM by employing the weak-beam technique as long as the dislocations are not too closely spaced.

4.5 Interfaces

4.5.1 Structure and Topography of Grain Boundaries

The more recent concepts concerning the structure of high angle grain boundaries, such as the coincident site lattice model (CSL model[163]) and the structural unit (SU) model[164] postulate the existence of *special low energy boundaries* of high periodicity with a certain proportion $1/\Sigma$ of lattice sites in the two grains forming a CSL, a sort of superlattice, which is continuous across the grain boundary[163]. According to the CSL model, small deviations from an ideal CSL-orientation are accommodated by either a single set of edge dislocations or a network of dislocations depending on the angle and axis of misorientation, and the inclination of the boundary plane[165, 166]; in general the Burgers vectors of the structural dislocations are not lattice vectors[166]. Furthermore, as has been pointed out by Brandon et al.[163], a grain boundary, which is constrained to lie at a certain angle (θ) to a close packed CSL plane, will minimize its energy by adopting a stepped topography in such a way as to maximize the total proportion of grain boundary area which coincides with a close packed CSL plane. The step height is related to the lattice parameter of the underlying CSL, i.e. to the value of Σ, and the number of steps depends on θ, although a boundary with many small steps might be energetically as favourable as one with a few large steps[163].

The structural unit model[164, 167] which is a modification of the CSL model and also postulates the existence of *special* boundaries, regards the interface between two crystals with a coincidence orientation relationship as consisting of identical segments which repeat periodically. In order to obtain minimum energy configuration the crystals in CSL orientation relationship may undergo a rigid translation with respect to each other before the individual atoms at the interface are able to relax[168, 169]. Hence, the structure of a (symmetrical) boundary between crystals with an ideal coincidence relationship is determined by the relaxed structure of a single unit, which is repeated periodically. If the orientation relationship is not an ideal coincidence one, the SU model predicts that the structure of the boundary is not drastically changed but rather is gradually modified with increasing misorientation to one of the neighbouring coincidence orientation relationships in adequate proportions[167].

It is quite natural, that the predictions from various theories concerning the structure and topography of high angle grain boundaries, have stimulated many experimental investigations using various techniques. Amongst these techniques field-ion microscopy and TEM have been employed most frequently since both techniques provide information about the crystallography of the boundary and its structure, e.g. grain-boundary dislocations and grain-boundary topography. Often FIM and TEM are used in combination in order to compensate for some inherent limitations from which both techniques suffer, if employed exclusively, as has been discussed recently by Smith[170] and by Loberg and Norden[171]. As has been pointed out in Chap. 2.5 and 4.1 most non-refractory metals, containing defects such as dislocations and grain boundaries, will rupture under FIM imaging conditions, and hence, these materials are not amenable to the study of grain boundary structures. Consequently, most FIM investigations of grain-boundary structures have been confined to iridium and primarily to tungsten.

There exist two further problems in studying grain-boundary structure and topography by FIM. Compared to the grain size of most materials, the volume of the FIM tip is rather small and, hence, the probability of finding a grain boundary in a FIM image is low. Although this problem may be to some extent circumvented by the special preparation technique for FIM specimens containing grain boundaries, which was described in Chap. 2.1, it probably accounts for the fact that many publications dealing with FIM studies of grain boundaries only report the analysis of a single boundary found more or less accidentally in the FIM image. In contrast, the much larger sampling volume accessible to TEM, provides a greater opportunity for systematic studies involving the selection of specific boundaries, e.g. CSL or near CSL boundaries with simple, periodic structures for which experimental and theoretical results may be compared.

In order to increase the probability of finding a grain boundary in a FIM tip, a small grain size is required; this is usually achieved by cold drawing, sometimes followed by slight anneal. The wires produced in this way generally exhibit a strong wire texture, with the wire axis being parallel to $\langle 110 \rangle$ in bcc tungsten. Due to this texture, the most frequently observed boundaries in tungsten are therefore $\langle 110 \rangle$ tilt boundaries. Thus, FIM studies of grain boundaries in tungsten are more or less confined to those $\langle 110 \rangle$ tilt boundaries which often contain the wire axis[171, 173]; sometimes a (small) deviation from pure tilt configuration is observed, caused by an additional rotation of only a few degrees about another axis.

Commonly, in a field-ion micrograph a grain boundary is observed as a dark band or a dark line (Fig. 4.10) indicating some preferential field evaporation of atoms in the boundary region. Because of this preferential field evaporation effect the apparent width of the dark band (about one to two atomic layers) is not necessarily identical with the actual grain-boundary width. However, all field-ion images of grain boundaries show the regular atomic structure to be maintained up to the dark band, thus giving direct evidence that the boundary region is not a thick liquid-like layer. Although the optimum (lateral) resolution of the FIM (about 0.3 nm) is usually achieved for tungsten (Chap. 2.7.2), it is still not sufficient to resolve the position and, hence, any displacements of atoms in the boundary region. Furthermore, as was pointed out in Chap. 2.2, not all atoms, but only those in kink or ledge site positions are imaged. Therefore, in practice it is *not* possible to verify the configuration of the structural units and the associated relaxation, which have been computer-modelled for various CSL orientation relationships[172], by FIM.

It has been shown by Smith et al.[174, 175] that the resolution normal to the surface is substantially better than parallel to the surface due to an *indirect magnification effect*. This phenomenon allows to detect in the FIM image any relative rigid translation of the two grains, if the associated displacement vector has a component greater than about 0.03 nm *normal* to the surface. However, due to the uncertainty in the determination of the component parallel to the surface, in general it will be not possible to interpret the rigid translation quantitatively[170].

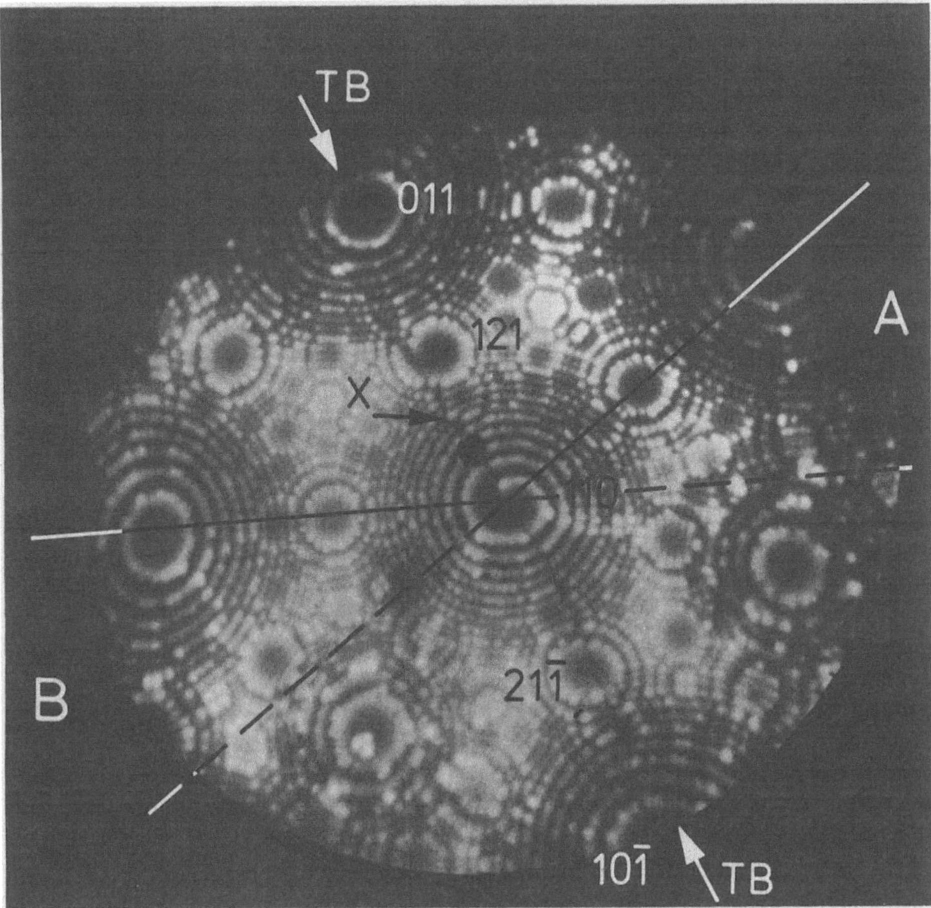

Fig. 4.10. Field-ion micrograph of an incoherent (111) twin boundary (TB) in tungsten. At X it is recognized that the boundary is not planar; also shown are some straight *zone lines* in both grains *A* and *B*. (Courtesy P. A. Beaven)

The procedures, which have been proposed in order to determine the crystallographic grain-boundary parameters, such as axis and angle of rotation and the indices of the boundary plane, from field-ion micrographs are usually based on the assumption that the tip has a cylindrically symmetrical endform with a spherical cap and, hence, that the field-ion image can be considered to be a stereographic projection of the tip (e.g. Ref. 10, 78, 176, 177). In this *idealized* case the trace of a *planar* boundary should be revealed either as a straight line if it runs through the centre of the field ion micrograph, or as a curved line if it intersects the tip surface away from the centre. This curved trace corresponds to a small circle in the stereographic projection, from which the boundary plane indices may easily be determined. However, in reality, the boundary is often neither planar, nor is the projection an ideal stereographic one[178, 179] due to the asymmetry in the shape of the field-ion specimen (see Chaps. 2.5 and 2.7.1).

Therefore, the boundary plane often does *not* correspond to a small circle in the stereogram and, as has been pointed out by Loberg and Norden[171], it is often not possible to decide whether this is due to variations in the projection, or to small changes in the boundary plane. Under these unfavourable circumstances the average boundary plane can only be determined within a few degrees[173]; in cases where the trace of the boundary is composed of several recognizable segments, the analysis can be performed for each segment. In the more favourable (and fortunately in bcc metals not uncommon) cases where the boundary plane contains the specimen axis (e.g. Fig. 4.10) the boundary plane can be determined more accurately.

It is evident from Fig. 4.10 that all zones passing through the centre of projection, i.e. through the (110) pole, are straight. In this case, and also in all cases where the boundary trace is close to the centre of projection, the angle and axis of rotation can be evaluated by extrapolating several straight zone lines from one grain into the other and assigning the indices of the points of intersection with respect to each of the two crystals[78, 170]. The misorientation can then be determined within about 0.5° following the computing method given by Smith and Goringe[177].

In order the predict the contrast pattern of a particular grain boundary in a FIM image, most contrast theories consider the matching of two sets of rings across the interface[176, 180, 181]; each set of rings is formed by a plane stack $(h_1k_1l_1)_A$ and $(h_2k_2l_2)_B$ in grain A and B, respectively, abutting across the interface. Depending upon the crystallographic parameters, such as the axis and angle of rotation, the indices (hkl) of the boundary plane, and the interplanar spacings $d\,(h_1k_1l_1)_A$ and $d\,(h_2k_2l_2)_B$, the contrast patterns from various types of boundary can be quite different from each other[176, 181]. This is illustrated schematically in Fig. 4.11 a–d for the special case when the trace of the boundary plane intersects the centre of a low index pole which is either common to both grains (Fig. 4.11 a without and 4.11 b with a rigid translation) or different for both grains (Fig. 4.11 c before, and Fig. 4.11 d after some relaxation). Fig. 4.11 e depicts a symmetrical tilt boundary with perfect *ring matching* between the poles of the same indices (hkl) in each grain, whereas in Fig. 4.11 f the more complex (though frequently observed) contrast pattern of a tilt boundary with a superimposed twist misorientation is shown; analogous to Fig. 4.11 d, the relaxed ring matching across the interface results in the terminating spiral AC whereas the twist misorientation is accommodated by the two dislocation spirals B and D.

In the context of ring matching, it should be pointed out that apparent perfect ring matching across the interface need *not* necessarily indicate that the atoms at the boundary occupy lattice sites, but again could account for the limited lateral resolution of the FIM which does not allow to detect small relaxations of atoms.

Figure 4.10 shows a tilt boundary in tungsten[182] resulting from a rotation of 70.5° about [110] which is identical with the wire axis. This rotation corresponds to a $\Sigma = 3$ coincidence orientation relationship and represents the special case of a twin orientation. The trace of the twin boundary intersects the specimen surface across the central (110) pole and passes through two further {110} poles, which are all common to both grains. Subsequent fine-scale field evaporation revealed initially no inclination of the interface and, hence, the boundary plane is defined as (111) with respect to both grains, and the twin boundary is therefore incoherent. Only after extended field evaporation the boundary plane deviated from (111). In Fig. 4.10 the rings across the tilt boundary plane match perfectly but all low index poles (e.g. 10Ī, 21Ī, the central (110) pole, 12Ī, and 011)

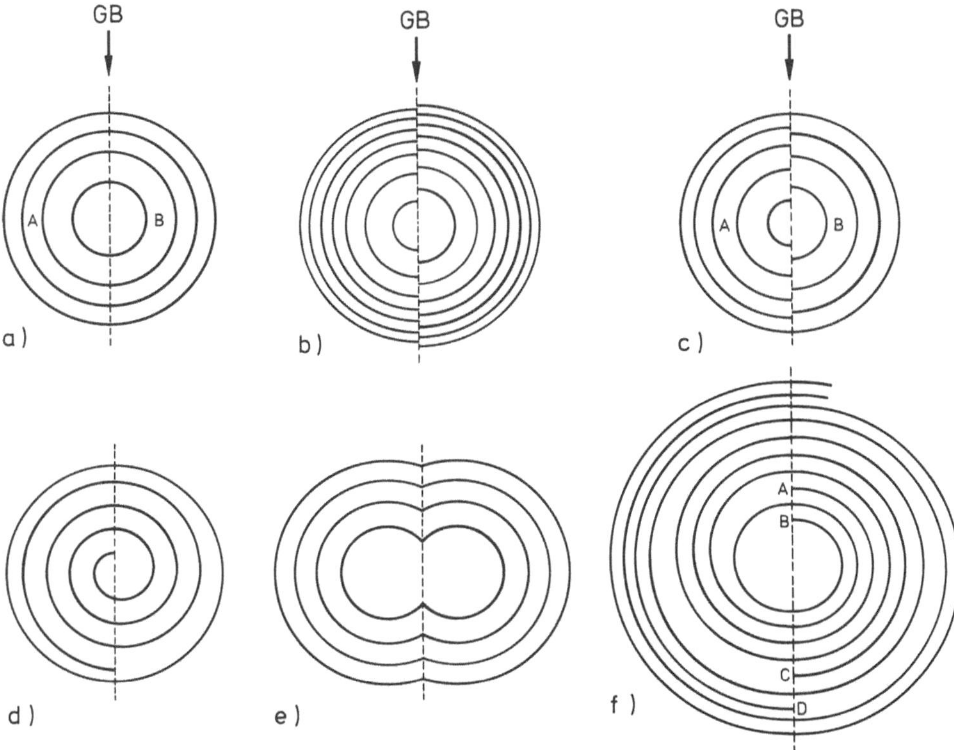

Fig. 4.11 a–f. Schematic drawings of some contrast patterns from various grain boundaries according to the ring-matching models. *GB* denotes the trace of the boundary plane. **a)** Perfect 1 : 1 ring matching of two low index poles common to grains *A* and *B* whose normals are identical with the axis of misorientation. **b)** The same after a rigid translation parallel to the rotation axis; the planes are now staggered and two sets of half rings appear. **c)** Two plane stacks of different lattice spacings meet at the boundary. In the unrelaxed case several non-matching half rings and a few matching rings will be observed; the number of the latter depends on the difference of the lattice spacing and on the angles between the normals of the two plane stacks and the boundary plane. **d)** The same as before after relaxation of the structure has occurred resulting in a terminating spiral. **e)** Ring matching across a symmetrical tilt boundary. In contrast to **a)** the normals of the two abutting poles are no longer identical with the axis of misorientation. **f)** See text. (After T. F. Page, P. R. Howell and B. Ralph[181])

exhibit spirals indicating the presence of several dislocations. During the course of fine-scale field evaporation the origins of the spirals moved along the boundary trace in a manner which suggested that the dislocations form a regular network in the boundary plane with a mesh size of about 10 nm.

Smith et al.[170, 174] pointed out, that in general it is not possible to determine the magnitude and the direction of the Burgers vector of a grain boundary dislocation in the FIM unless it intersects the tip surface at different poles during the course of field evaporation. In the very special case of the above mentioned regular network this requirement is met and, hence, Beaven et al. have been able to conclude from their FIM studies that the small twist misorientation of the incoherent twin boundary is accommodated by a hexagonal network of screw dislocations with Burgers vectors of the type $a/3$ $\langle 112 \rangle$; this conclusion has been corroborated subsequently by TEM. These Burgers vectors do not correspond to lattice vectors but rather are primitive vectors of the corresponding DSC lattice in agreement with theoretical predictions[166]. Furthermore, these results have also shown that the grain-boundary defect structure is not affected by the high electric field stress acting on a FIM specimen.

Similar FIM observations of coincidence related boundaries with $\Sigma = 3, 11, 17, 19, 33$ (e.g.[163, 171, 173, 183–185] and even $\Sigma = 41$[186] together with a superimposed network of dislocations have been reported quite frequently; in most studies the boundaries approximated to tilt boundaries with rotations about the $\langle 110 \rangle$ tip axis. However, in none of these FIM investigations could the grain-boundary structure be completely analysed without additional information from complementary TEM studies[171, 187].

So far, none of the numerous FIM studies of CSL boundaries has revealed a contrast pattern similar to that shown schematically in Fig. 4.11 b, which is produced by a relative translation of the grains normal to the surface. This fact may indicate that either there are no rigid translations associated with the analysed boundaries, or that the displacement is less than 0.03 nm and, hence, not detectable, or there occurs bulk relaxation, as suggested by Loberg and Norden[171], resulting in the frequently observed pattern shown in Fig. 4.11 a. The experimental verification of rigid translations associated with CSL boundaries in aluminium and stainless steel by means of TEM[188, 189], which is able to detect displacements as small as 4×10^{-3} nm under special circumstances, points towards the second explanation; however, further clarification of this point should be obtained by analysing several CSL boundaries by both FIM and TEM.

From the observation of networks of structural dislocations, which accommodate small deviations from a coincidence relationship, it was deduced that CSL boundaries have energies lower than those of random boundaries. In order to check whether the[110] tilt boundaries most frequently observed in FIM investigations show a preference for special orientations according to the CSL model, Loberg et al.[173] collected data from various FIM investigations and plotted the angles of rotation about a $\langle 110 \rangle$ axis (within $3°$) and the indices (hkl) of the corresponding (mean) boundary planes into one diagram (Fig. 4.12). The full squares represent the special misorientations θ_{110} and the corresponding special boundary planes as predicted by the CSL model for values of Σ up to 51; the open circles show the experimental data and, in order to refer crystallographically to both grains, each asymmetrical boundary is represented by two data points.

It is evident that the misorientations are often near to those expected from the CSL model, but there is no tendency for the data points to cluster around misorientations of low Σ, and indeed, more recent FIM investigations by Ishida and Smith[186] have provided further evidence for the significance of special boundaries with large Σ, e.g. $\Sigma = 41$.

As is recognized from Fig. 4.12, the experimental values of θ_{110} deviate from the *special* misorientations often by a few degrees which is definitely no longer within the experimental uncertainty (about $0.5°$). At this stage the question of the range of applicability of the *CSL plus dislocation-network model* arises, i.e. whether these large deviations from an ideal CSL orientation relationship can still be accommodated by a network of structural dislocations (for discussion see[190]). Answering this question necessitates the resolution of individual grain boundary dislocations whose spacing at larger deviation angles may be too small to be resolved in the TEM and the necessary information may be obtained only from FIM studies.

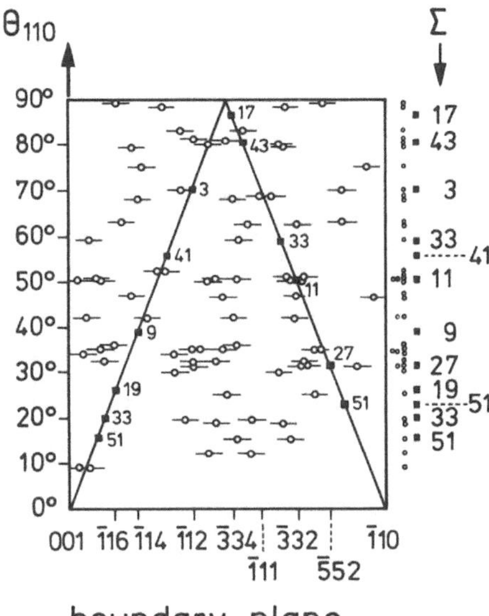

Fig. 4.12. A collection of observed misorientations and indices of corresponding grain-boundary planes for [110] tilt boundaries. (After B. Loberg, H. Nordén and D. A. Smith[173])

According to Warrington[191], only about 9% of all boundaries in a polycrystal, with its grains *randomly oriented* with respect to their neighbours, are covered by CSLs (with Σ between 3 and 25) *plus* dislocation networks. Although this small value of coverage might be significantly increased by the presence of a distinct texture as is the case in most FIM specimens, it could well be argued that the scatter of observed misorientations in Fig. 4.12 provides some evidence for a small proportion of special boundaries in a polycrystal.

Most FIM investigations dealing with the topography of grain boundaries report the observations of non-planar boundaries often containing ledges and small protrusions of only a few nm similar to the one observed (at X) in the trace of the incoherent twin boundary of Fig. 4.10. Brandon et al.[163] observed a high density of ledges at a Σ = 11 boundary, the average boundary plane of which was at an angle of about 9° to the most densely packed plane of the Σ = 11 coincidence lattice. Although they did not perform any crystallographic analyses of the facets and ledges of the boundary, the existence of the ledges has been taken as a direct confirmation of the theoretically predicted topography. Morgan and Ralph[192] analysed a grain boundary in cold-worked iron characterized by a rotation of 31° about a [230] axis; the boundary was macroscopically curved but could be divided into three planar sections each of which was parallel to the specimen axis. It was concluded that each planar section coincided with a more densely (though not *most* densely) packed plane of the CSL lattice. This conclusion, which has been taken as some further support for the CSL model, has been criticized by Loberg and Nordén[171], who showed that the misorientation of 31° about [230] yields a CSL of Σ = 91 instead of Σ = 7 as erroneously assumed by Morgan and Ralph.

More details about the boundary topography on a small scale can be obtained by field evaporation along the trace of a boundary using small evaporation increments. From the recorded points along the boundary trace the topography is then reconstructed either isometrically[193] or by drawing contour maps, in which deviations within 0.5 to 1 nm from the *average* boundary plane can be detected[78, 194, 195]. Figure 4.13 shows the topography

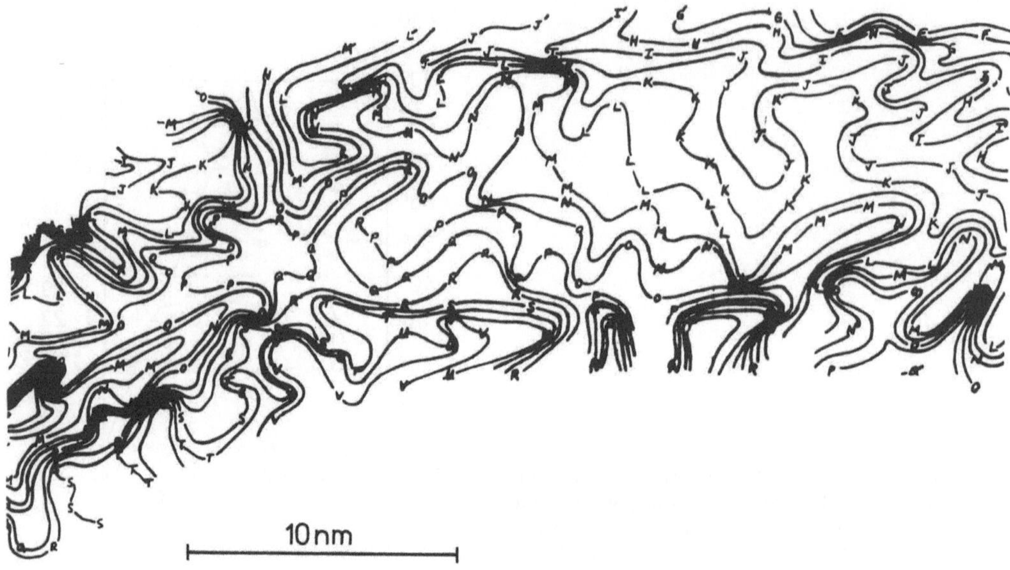

Fig. 4.13. Topography of a grain boundary in a dilute *Mo*-Ti alloy. The letters at each contour indicate the height of the boundary surface with respect to the mean boundary plane (labelled *K*). Letters *A* to *J* and *L* to *Z* represent succesive 0.4 nm contours below and above this plane, respectively. (After T. J. Wilkes, G. D. W. Smith and D. A. Smith[78])

of a boundary surface in a dilute as-drawn *Mo*-Ti alloy (27° about [20, 19, 1], thus not corresponding to a coincidence misorientation) as reconstructed in the form of a contour map by Wilkes et al.[78]. Some relatively planar segments extending sometimes over more than 5 nm are observed together with several sharp protrusions the height of which reaches in some instances 4.5 nm with respect to the mean boundary plane. In general, the complex nature of the interface containing small ridges, valleys and protrusions has been confirmed by Bolin et al.[195] in a similar study of various boundaries in tungsten. In one case, they showed that the topography of a CSL boundary with $\Sigma = 11$ is composed of several planar segments fluctuating approximately ± 1 nm about the (001) plane, thus forming a *crinkled* surface which still approximated a (001)$_A$ plane on a large scale (about 100 nm^2). Construction of the $\Sigma = 11$ CSL revealed that by adopting a crinkled surface, the number of coincidence sites in the boundary is significantly increased as compared to a planar (001) boundary plane. Whereas this observations is in accordance with the CSL model, Bolin et al. showed also that several other boundary planes of grain boundaries, even in coincidence misorientation, did not follow the predictions of the CSL theories; this however, is also evident from inspection of Fig. 4.12.

In conclusion to the above section, it may be stated that from the microscopic point of view the boundary surface is not planar but rather complex; furthermore, the topography of most grain boundaries analysed by FIM does not follow the lines of the CSL model. One possible explanation for this fact might be that CSL theories deal with grain boundaries in minimum energy configurations whereas by FIM commonly only boundaries in cold-worked materials have been investigated; these boundaries, however, may not have yet adopted the energetically most favourable topographical configuration. Some evidence for such a explanation has been given by Howell and Ralph[196] who reported the topographical structure of a boundary (44° about [110]) in *recrystallized* tungsten to be less complex than that in deformed specimens; however, again, the ledges and steps observed in the boundary trace did not correspond to those predicted by CSL theories.

4.5.2 Interphase Interfaces

A contrast theory, based on a ring matching approach similar to that discussed in the preceding chapter (Fig. 4.11), has been developed by Hildon et al.[197] in order to determine interfacial parameters, such as orientation relationship, the state of coherency, and the topographical structure, of interphase boundaries in two-phase materials. Although this theory has been successfully applied to a few interphase interfaces[198], its range of applicability is confined to (rare) cases where *both* abutting phases exhibit a rather well developed ring structure. Especially in precipitating systems, where the interfacial parameters are often of particular interest, this requirement is frequently not satisfied due to the difference in both the best image voltage and the field evaporation characteristic of the two phases (see Chap. 2.6.3). For this reason, only few quantitative FIM studies have been reported concerning the nature of interphase interfaces. In two of these, Hildon et al.[198] and Wilkes et al.[199] reconstructed in form of contour maps the topography of interphase boundaries between a Co_2Ta precipitate and the cobalt matrix, and between a TiN platelet and the molybdenum matrix, respectively. Consistently, both studies reported a complex heavily ledged interphase interface. The contour plot of the interface between a γ'-Ni_3Al particle and the α-Ni Matrix also revealed a ledged structure[200] indicating that coarsening of Ni_3Al precipitates occurs via a ledge mechanism rather than by *direct* volume diffusion[201].

4.5.3 Interface Segregation

Due to its excellent spatial and depth resolution in combination with the facility for identifying chemically any impurities in a quantitative manner, the atom-probe FIM is ideally suited to the study of impurity segregation at interfaces. Furthermore, with this technique it is possible to study segregation phenomena in-situ on an *unfractured* grain boundary, whose type and structure can be characterized *simultaneously* in much detail.

The imaging behaviour in the FIM of most interstitial segregants is analogous to that of interstitial impurities in the bulk material (see Chap. 4.2.3). For example, oxygen, the most frequently identified interstitial segregant shows up as bright spots along, and in, the proximity of the boundary[202–204]. A comparison of the number of bright spots along the grain boundary in tungsten *oxygenated* at 1400 °C with that in an as-received specimen clearly established that oxygen atoms segregate to the boundary, and that they appear as bright spots. However, again there exists no one to one correspondence between the abundance of bright spots and the number of oxygen atoms; furthermore, it is not possible to distinguish between different segregants by just observing bright spots. Therefore, the information about grain boundary segregation obtained from FIM studies, which has been essentially based on bright spot counting, can only be of a qualitative character and any absolute figures given must be regarded with caution. Fortes and Ralph[203] reported that they only found marked segregation of oxygen at grain boundaries in iridium if the boundaries were *not* in a CSL orientation; in this case the segregated zone extended to about 45 nm ± 5 nm on each side of the interface, the peak concentration at the boundary being six times higher than the matrix value.

Turner and Papazian[205] used the atom-probe FIM in order to analyse quantitatively the segregation of carbon at a grain boundary in a Fe-0.05 at % C alloy annealed at 723 K. They also found a broad segregation zone (about 50 nm) exceeding 0.4 at % of carbon at the interface. This peak value, however, is still much less than one would expect if the boundary were to be covered with a carbon monolayer.

During the past few years, similar, though more comprehensive, studies af various steels, having undergone various thermo-mechanical treatments prior to atom-probe microanalyses, have also reported a marked segregation of carbon to grain boundaries and to coherent twin boundaries[206–208]. The segregation of substitutional alloying elements, such as Cr and Si, to grain boundaries in stainless steel[207] and in manganese steels with different additions of silicon has also been investigated. In the latter steel, Mintz and Turner[210] observed that in the high-Si steel, silicon segregated at the grain boundaries whereas in the low-Si steel an enrichment of chromium was found at the interfaces; simultaneously it was shown, that a larger addition of Si decreased the tendency of interstitial impurities (C, N) to segregate at grain boundaries.

The transformation behaviour of steel, which is largely influenced by alloying constituents, such as Cr, Mo, Mn, etc., has been the subject of many research activities during recent years. In this context, much effort has been directed towards the analysis of the distribution of these elements and of carbon in the transformation products, e.g. martensite, pearlite or bainite. For this type of analysis the atom probe is particularly powerful as several investigations have demonstrated (e.g. see[211]). A common feature of these studies is that they all report significant changes in the distribution of the alloying species across the interphase boundaries, such as the cementite – ferrite interphase in a pearlitic steel[212, 213] or the ferrite – austenite/martensite interface[214] in a bainitic CrMo steel.

Miller et al.[213] and Williams et al.[212] investigated the distribution of Cr, Mn and Si in both a cold-drawn and in a transformed patented pearlite steel. It was found that redistribution of the alloying elements between the cementite and ferrite phases takes place, even at temperatures well below the thermodynamically predicted no-partition temperature. By determining concentration profiles across the ferrite and cementite phases, it was shown (Fig. 4.14) that the cementite-ferrite interface is significantly enriched in Cr, Mn and Si with respect to both phases. In order to explain the form of the concentration profiles after a given transformation time, diffusion calculations were carried out, and it was concluded that considerable redistribution of chromium occurs via a short-circuit diffusion path in the transformation interface.

In a recent study, Bach et al.[206, 207] found some evidence that molybdenum segregates to the ferrite-austenite interface during the transformation of a low-alloyed CrMo steel from austenite to bainite. Although such a segregation of Mo has been proposed to explain the observed retardation of the transformation if Mo is added as alloying constituent[215], the results still have to be corroborated by additional experiments. The same holds true for a preliminary study of the trapping of ion-implanted antimony in the

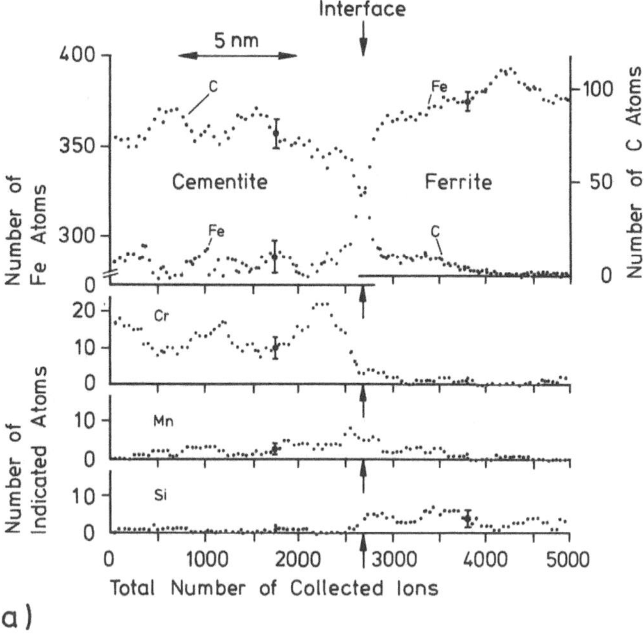

a)

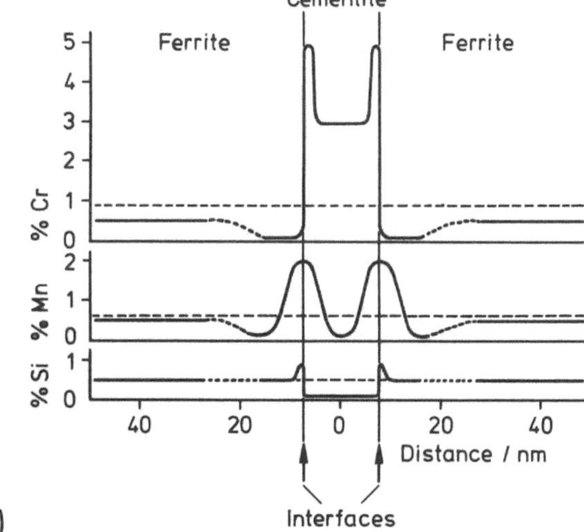

b)

Fig. 4.14. Composition profile for various elements across a cementite-ferrite interface as determined with the atom probe. Each data point represents the analysis of 400 ions. **b)** Schematic distribution of the alloying constituents in the vicinity of cementite-ferrite interfaces as concluded from the atom probe analyses. (After P. R. Williams et al.[212])

interphase boundary region between TiC particles and ferrite after annealing at 550 °C[216].

Future studies of segregation by means of the atom-probe FIM will probably also utilize the imaging atom probe (IAP) as a complementary technique. Due to the considerably larger area sampled by the IAP (see Chap. 3.5) it is possible, for instance in finely dispersed precipitating systems, to analyse segregation at several interphase boundaries *simultaneously*. However, since the IAP does not (yet) provide the concentration of segregants fully quantitatively, this information still has to be obtained using the conventional ToF atom probe. Combined use of both instruments will yield more information about the segregation to interfaces, than can currently be obtained by any other technique. This has been demonstrated recently by Waugh et al.[217] in an investigation of the carbon redistribution in heavily drawn pearlitic steel after an anneal at about 200 °C.

The beauty and power of the IAP for the study of segregation at interfaces has been illustrated, also by Waugh et al.[4], in analysing the segregation of oxygen at a grain boundary in molybdenum. In the left part of Fig. 4.15, the grain boundary at which segregation of oxygen has occurred, is clearly seen in the field-ion image (arrowed); the position of the grain boundary is also evident in the corresponding Mo^{3+} image (centre part of Fig. 4.15) which was obtained by field evaporation of some atomic layers and gating for the Mo^{3+}-ions (Chap. 3.5). To obtain the distribution of oxygen, the imaging atom probe was gated for O^+ ions and a total of thirty atomic layers was removed (right part of Fig. 4.15). As is clearly to be seen, the oxygen atoms are segregated within a ~ 1 nm wide band along the grain boundary; areas further away from the grain boundary, only show a small density of bright spots, i.e. the oxygen concentration in the bulk is fairly small.

4.6 Precipitation

4.6.1 Some General Remarks

Many technologically important properties of alloys, e.g. the mechanical strength, are essentially controlled by the presence of precipitated particles of a second phase which are formed in the originally supersaturated matrix. A fundamental understanding of the thermodynamics, the kinetics and, hence, the mechanism of precipitation reactions in metallic solids, resulting in a well defined microstructure, is therefore of great interest in materials science.

The course of a precipitation reaction, including the early stage decomposition as well as the coarsening stages or any transition stages such as occur in a complex precipitation sequence (see Chap. 4.6.2) can, in general, not be followed continuously by any experimental technique. Therefore, the progress of a precipitation reaction is reconstructed from the microstructure which develops at various stages during the phase transformation. In essence, the microstructure is characterized by the distribution, volume density, size, morphology, and chemical composition of the precipitated particles.

The analysis of microstructures resulting from precipitation, therefore, requires microanalytical techniques which are capable of resolving very small particles and of identifying simultaneously their chemical composition. Whereas the first requirement is met by TEM (at least for most particles larger than about 4 nm) and under some circumstances also by neutron- und X-ray small angle scattering techniques, the *chemical analysis* of particles *smaller* than about 12 nm is in general not possible with any of the standard microanalytical tools, such as electron microprobe or Scanning TEM combined with an energy-dispersive X-ray analyzer. Therefore, many precipitation structures of technically important materials composed of very small, finely dispersed, precipitates, e.g. G. P. zones or carbides, are not amenable to analysis with any of the above mentioned microanalytical tools. It is especially in this area, where the FIM combined with an atom probe has provided useful information as will be discussed in the following chap-

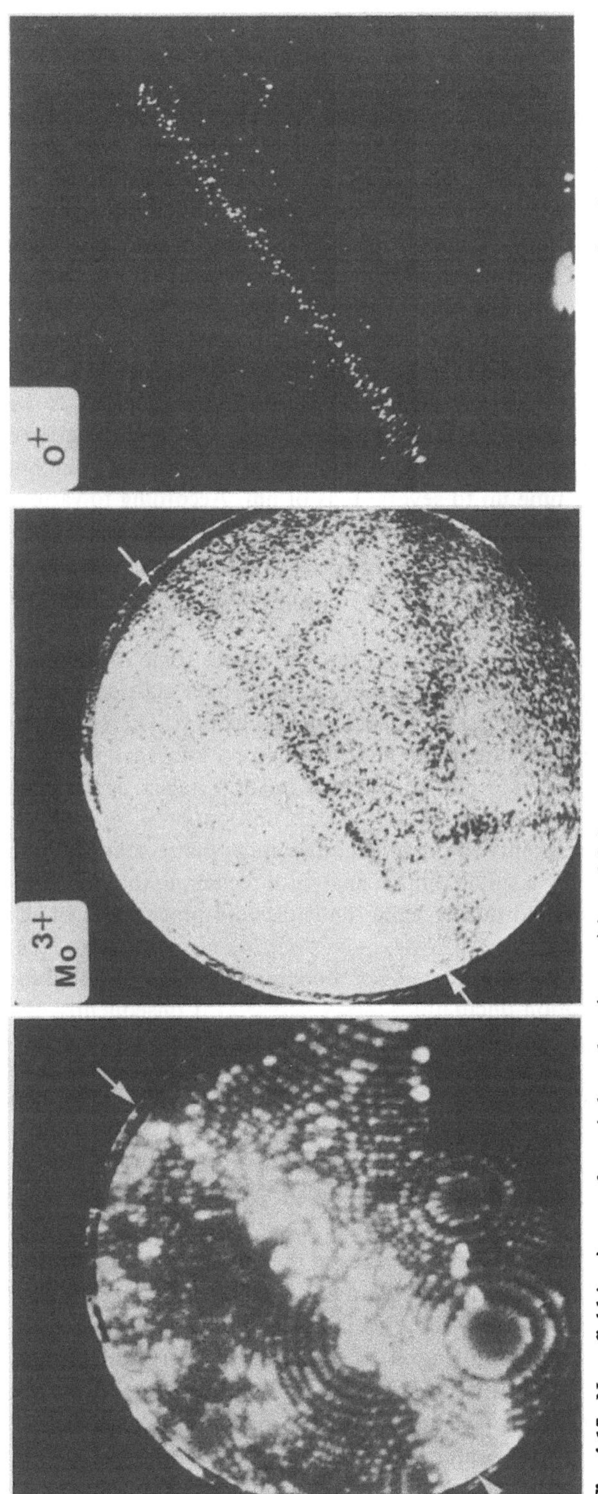

Fig. 4.15. Neon field-ion image of a grain boundary (*arrows*) in molybdenum (*left*); the position of the boundary is also visible in the time-gated Mo^{3+}-image (*middle*). The time-gated O^+-image shows a considerable number of oxygen atoms segregated to the grain boundary (*right*). (Courtesy A. R. Waugh)

ters. However, since much work has been done in this field it is not possible to give an account of all reported studies but rather to review a small selection of some more comprehensive investigations.

Due to the lack of sufficient resolution, it has also not been possible with one of the above mentioned conventional microanalytical techniques to study the *very early stages* of decomposition in a supersaturated alloy. Therefore, for many two-phase alloys the theoretically important question of whether the phase transformation was initiated by the *formation of small nuclei* having already a composition similar to that of the equilibrium phase (or, at least, similar to that of the intermediate metastable phase which is formed in some alloys prior to the formation of the equilibrium phase) or whether it was initiated by *spinodal decomposition* remains open. Since the latter reaction is caused by a thermodynamic instability of the quasi-uniform solid solution within the spinodal range [e.g. Ref. 218], the decomposition occurs without any nucleation barrier larger than kT to be overcome, by the growth of three-dimensional composition waves. The wavelengths of these long-range composition fluctuations are initially typically of the order of 5 nm but will subsequently grow with aging time up to several tens of nm. According to various theories of spinodal decomposition[219, 220], the amplitudes of the composition waves will – in contrast to nucleus formation – increase gradually with aging time starting from almost zero until they reach a value which corresponds to the concentration of the completely precipitated second phase.

The microstructure in the *later* transformation stages (i.e. where some of the mentioned microanalytical tools can now be applied), evolving from either classical nucleation or spinodal decomposition, do not necessarily differ from each other. Therefore, it is in general not justified to deduce the mode of the initial decomposition reaction simply from observations of the microstructure in the later transformation stages, as has been done quite frequently in the past.

To our knowledge, the only possibility to prove experimentally that a solid solution decomposes spinodally is to show by a suitable microanalytical technique that the composition waves are gradually amplified during aging until discrete precipitates of the second phase have finally emerged.

The atom-probe FIM meets all the requirements of a microanalytical tool suited for the analysis of long-range composition fluctuations (see Chap. 3.2). Consequently, the technique has been applied (though only in very recent years) to study the phase transformation in a few supersaturated alloys which are thought to decompose spinodally; some results and conclusions obtained from some of these investigations will also be discussed.

4.6.2 Phase Separation by Nucleation and Growth

One of the first comprehensive atom-probe FIM studies of precipitation was performed by Goodman et al.[221] in a Fe-1.4 at% Cu alloy isothermally aged at 500 °C. In this alloy system, the Cu-rich precipitates are visible in the FIM image in dark contrast as long as their diameter (2r) is less than about 5 nm (Fig. 4.16a); particles of larger sizes show some structural details (Fig. 4.16b) but they can still be easily distinguished from the matrix. From the matching of the lattice plane edges from the matrix into the particles it was concluded that particles larger than 5 nm are no longer coherent with the matrix. Due to the visibility of the precipitated particles in the field-ion image, the diameter of

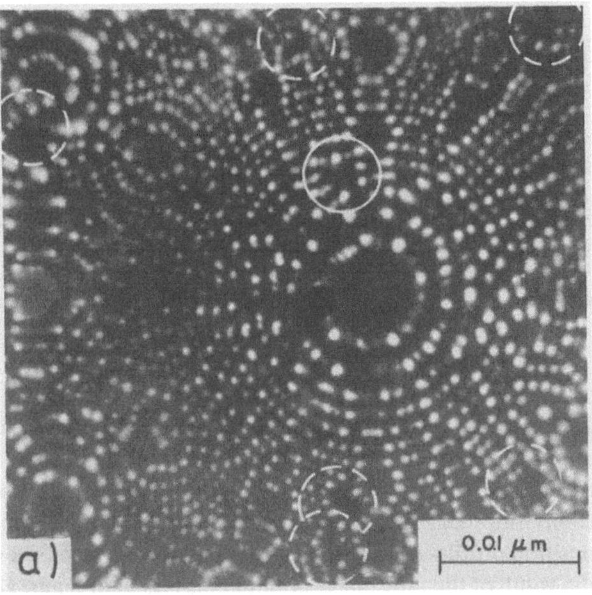

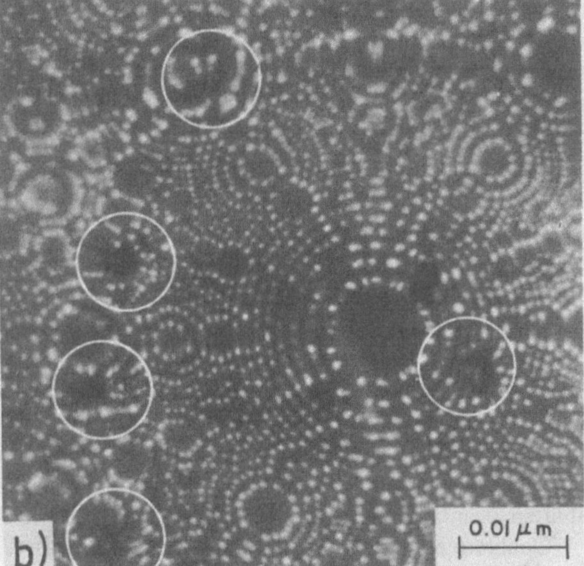

Fig. 4.16 a, b. Cu-rich precipitate particle in aged Fe-1.4 at% Cu. **a)** 1 h at 500 °C; particle within *solid circle* was analysed with the atom probe (Fig. 4.17) **b)** 9 h at 500 °C. (Courtesy S. S. Brenner)

single particles as small as 0.8 nm could be determined quite accurately (within about 0.2 nm) by employing the "persistence-size technique" outlined in Chap. 2.7.1 (in the TEM only particles larger than 5 nm in diameter became visible). Furthermore, the composition of these small particles could be determined by placing the precipitated particle under investigation under the probe hole of the atom-probe. By this procedure, Goodman et al. showed that the Cu content of the precipitates continuously increased

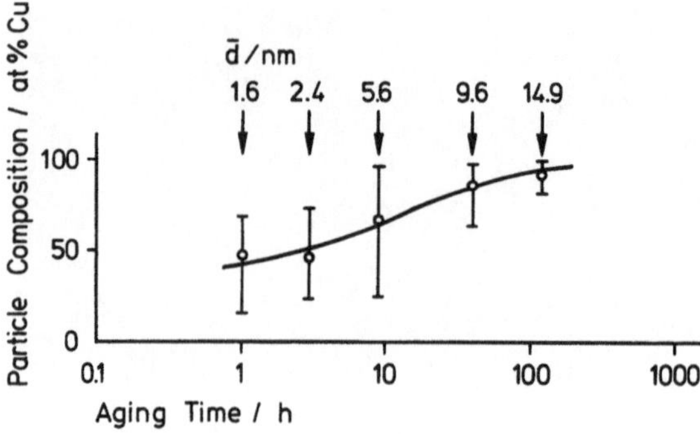

Fig. 4.17. Change of the mean composition and diameter d of particles in Fe-1.4 at % Cu with aging time at 500 °C. *Brackets* indicate ranges of composition. (After S. R. Goodman, S. S. Brenner and J. R. Low[221])

during aging for times up to 100 h (Fig. 4.17); only after the particles have grown to about 14 nm does the Cu concentration in the precipitated phase approach its equilibrium value of almost 100%. (It should, however, be pointed out that the scatter of the composition is still rather large even for the larger precipitates.) By analysing the composition profile across the particle/matrix interface (see Chap. 3.2) it was concluded that the width of the interphase boundary is less than 1 nm. During the course of precipitation the particle density decreases from about $10^{18}\,\mathrm{cm}^{-3}$ after 1 h to $9 \times 10^{15}\,\mathrm{cm}^{-3}$ after 120 h and only then did the precipitated volume fraction reach its equilibrium value. These results indicate that the nucleation period at 500 °C is terminated within less than 1 h; subsequently some existing particles start to grow at the expense of (smaller) dissolving particles and especially by further uptake of Cu atoms from the still supersaturated matrix.

Youle and Ralph[222] studied the decomposition of a very similar alloy (Fe-1.5 at% Cu) at a lower aging temperature (470 °C) utilizing only FIM. In contrast to Goodman et al., they reported the *mean* diameter $(2\bar{r})$ of the precipitates to remain constant $(\sim 2.9\text{ nm})$ for aging periods up to 5 h, whereas the number density of precipitates was found to increase from $1 \times 10^{16}\,\mathrm{cm}^{-3}$ after 0.5 h up to $1 \times 10^{18}\,\mathrm{cm}^{-3}$ after ~ 6 h before it decreased again $(5 \times 10^{17}\,\mathrm{cm}^{-3}$ after 12 h). The initial increase in the number density has been explained by assuming that during early stage decomposition, zones or clusters fairly low in Cu content are formed, and that these only image as particles when they have become rich enough in copper; this is supposed not to occur until their size has reached about 3 nm. This interpretation, however, should be further corroborated by recording concentration profiles with the atom probe; a subsequent autocorrelation analysis (see Chap. 3.2) should clearly reveal the presence of the (assumed) Cu zones.

Qualitatively the results obtained by Wendt and Wagner[75] from an investigation of early stage decomposition in Cu-1 at% Fe isothermally aged at 500 °C, are quite similar to those of Youle and Ralph; as can be seen from Fig. 4.18 the number density (N_v) also goes through a maximum during aging, the initial increase of N_v resulting from further nucleation of iron-rich precipitates. The mean particle size remains constant until the precipitation is completed; only then does coarsening seem to commence accompanied by a decrease in N_v. However, there remains some uncertainty over whether the constancy of the mean particle size is of real physical significance or whether it is simply an artifact caused by averaging over a rather broad distribution of cluster and particle sizes extending between about 0.5 and 2.4 nm; the size of the critical nucleus is expected to lie

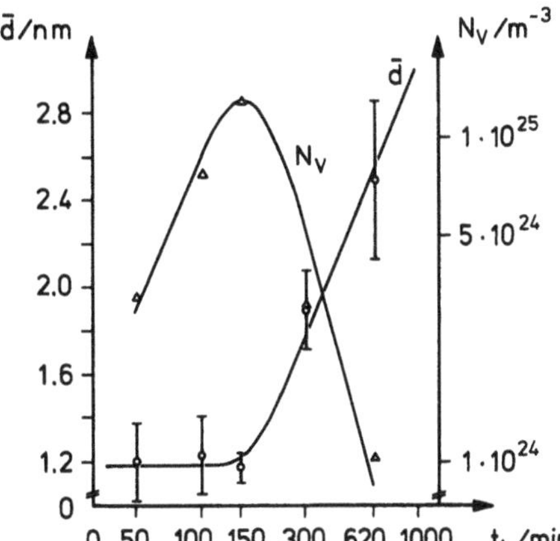

Fig. 4.18. Change of mean particle size d and number density N_v in Cu-1 at % Fe as a function of aging time at 500 °C. (After H. Wendt[75])

somewhere in between, but can not yet be evaluated due to the absence of necessary thermodynamical data. In this particular alloy, atom-probe analyses could not be performed because of the premature failure of the specimens whenever a field-evaporation pulse was applied; for this reason, no conclusions could be drawn concerning the composition of the visible Fe-rich precipitates (see Fig. 2.15).

Faulkner and Ralph[200] measured the coarsening kinetics of the precipitated γ' phase (Ni_3Al) in Ni-6.5 w% Al during aging between 650 and 740 °C utilizing FIM. In this two-phase system the γ'-particles only become visible if the specimen is either initially field-etched in hydrogen which is added at a low partial pressure ($\sim 1.4 \times 10^{-5}$ mb) to the neon imaging gas or if the specimen fractured (see Chap. 2.6.3). The sizes of about 25 γ'-particles were determined by the persistence size and/or trace analysis technique[223] for each heat treatment. Following the coarsening of the γ' precipitates with aging time (t_A) and analysing the data according to the Wagner-Livshitz theory[224] for diffusion controlled coarsening ($\bar{r}^3 \sim t_A$), yielded values for both the γ'/matrix interfacial energy, and the diffusivity of Al in Ni, which agreed quite well with values previously determined in a similar study employing TEM[225]. From this TEM work it was concluded that the early stage precipitation occurs via *nucleation* and subsequent growth of (spherical) γ'-precipitates, and that only during the later coarsening stages do the Ni_3Al-particles align themselves quasi-periodically (thus forming a *modulated structure*) in the matrix due to *selective growth*. In principle, the FIM results obtained by Faulkner and Ralph also agree in this respect, although they showed that even particles of less than 7 nm are quasi-periodically aligned and therefore form a modulated structure already in the early stages of aging.

Recently, Hill and Ralph[226] investigated once more the phase separation in Ni-14 at% Al at 625 °C, this time, however, by measuring composition profiles along $\langle 111 \rangle$ with the atom-probe FIM; the recorded composition profiles have subsequently been subjected to discrete fast Fourier transformations (see Chap. 3.2).

An analysis of these Fourier spectra yielded a few (*selected*) Fourier components with wavenumbers β ranging between 0.34 and 1.1 nm⁻¹ which grow with aging time whereas some others (again selected) having larger wavenumbers decay with time. Application of Cahn's theory of spinodal decomposition[219] to the time evolution of single Fourier components resulted in a *negative* interdiffusion coefficient which is a necessary and sufficient condition for spinodal decomposition. Hence, Hill and Ralph concluded – in contrast to the earlier FIM results obtained by Faulkner and Ralph – that phase separation in NiAl takes place by a *spinodal reaction.*

This conclusion, however, is disputable for at least two reasons: i) According to Cahn's theory *all* Fourier components with wavenumbers between zero and a critical value β_c (about 1.1 nm^{-1} in the paper of Hill and Ralph) should grow with time and *all* components with $\beta > \beta_c$ should decay with time. A close examination of the Fourier spectra of Hill and Ralph, however, reveals several Fourier components with $\beta \lesssim 1.1$ nm^{-1} which exhibit a decay in their amplitudes during aging; on the other hand, several components in the Fourier spectra with $\beta > \beta_c$ grow with aging time. Therefore, it is only through the arbitrary selection of some particular Fourier components in the spectra that it became possible to define a critical wavenumber β_c separating two ranges of wavenumbers whose associated Fourier amplitudes grow ($\beta < \beta_c$) or decay ($\beta > \beta_c$) with aging time, respectively. ii) The composition profiles of the specimens aged for the shortest period of this study (5 h at 625 °C) already clearly reveal the presence of discrete γ'-Ni$_3$Al precipitates indicating that phase separation had been nearly completed before the investigation was even started; hence, Hill and Ralph studied the coarsening stage of the γ'-particles rather than spinodal decomposition, i.e. early stage precipitation. (Even if one still were in the *late* stages of spinodal decomposition after aging for 5 h, it is no longer justified to analyse the Fourier spectra in terms of Cahn's linear theory).

The above criticism of Hill and Ralph's evaluation and interpretation of the atom-probe data has recently been corroborated by Wendt[53, 91]. The concentration profiles as determined by the atom-probe FIM revealed γ' particles already after aging for only 10 min at 650 °C. A subsequent analysis of the concentration profiles in terms of Fourier transforms and autocorrelation (Chap. 3.2) yielded the average particle diameter already to be 4 nm. In Fig. 4.19 the autocorrelogram and the corresponding Fourier spectrum of a specimen aged for 50 min at 650 °C are shown. It is obvious that the interparticle distance $\lambda \approx 18$ nm as determined from the autocorrelogram (peak at k_1 in Fig. 4.19 a) manifests itself as the *dominant* peak in the Fourier spectrum (arrow in Fig. 4.19 b) at a Fourier wavenumber $\beta_m \approx 0.06$ nm^{-1} corresponding to about k_1^{-1}. Due to the rapid phase transformation at temperatures above 650 °C it seems rather difficult to obtain information about the identity of the early stage transformation mechanism in Ni-Al and, consequently, Wendt reduced the aging temperature to 550 °C. By comparing autocorrelograms, which were computer generated by assuming various conceivable stages of phase separation[90, 91], with those experimentally determined, Wendt not only gained information about the (mean) particle size and interparticle distance, but also about the number density and the composition of the precipitates. He concluded that even the smallest particles ($\overline{D} \approx 2.0$ nm, Fig. 4.19 c) obtained after aging for only 10 min at 550 °C already contain at least 22 at% Al, as expected from the phase diagram for the γ' precipitates. Furthermore, the Fourier spectra of concentration profiles recorded after five different aging periods between 10 and 1400 min (in contrast, Hill and Ralph only recorded two profiles after aging for 300 and 3000 min at 625 °C) did not show any characteristic growth of Fourier amplitudes which could have been interpreted in terms of spinodal decomposition. Hence, altogether there is now overwhelming experimental evidence that Ni-Al decomposes via classical nucleation and subsequent growth, and *not* by a spinodal reaction.

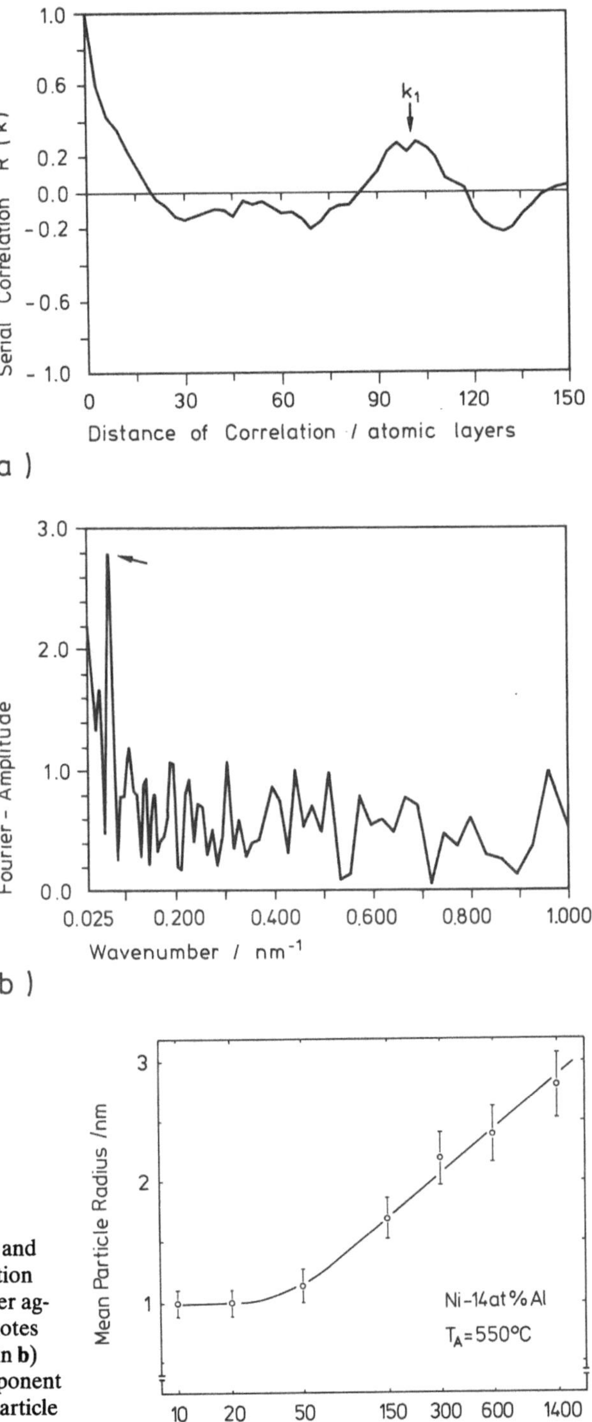

Fig. 4.19 a–c. Autocorrelogram (**a**) and Fourier spectrum (**b**) of a concentration profile recorded in Ni-14 at% Al after aging for 50 min at 650 °C. In **a**) k_1 denotes the interparticle spacing; the *arrow* in **b**) indicates the dominant Fourier component which also corresponds to the interparticle spacing. **c**) The variation of the mean radius of γ'-precipitates with time in Ni-14 at% aged at 550 °C. (After H. Wendt[91])

Beaven et al.[227, 228] employed the atom-probe FIM in order to investigate the distribution of alloying in a cast commercial nickel-based superalloy (containing Ci, Cr, Co, Ti, Al as major constituents) after a four stage heat treatment. This heat treatment resulted in a microstructure consisting of a dispersion of large ordered γ' particles ($\overline{D} \approx$ 135 nm) and a dispersion of finer secondary precipitates ($\overline{D} \approx 10$ nm), the latter could only be identified in the FIM (Fig. 4.20) but not in the TEM. Although this study was considered to be preliminary, it showed that both the γ' precipitates and the secondary precipitates correspond to the stoichiometric composition $(Ni, Co)_3 (Al, Ti, Cr)$. However, whereas the primary γ' precipitates contained 13.4 ± 1.2 at% Co the analyses of secondary γ' particles, which only form after the last stage of the four-stage heat treatment, yielded only 3.7 ± 0.7 at% cobalt.

The FIM has also been applied to investigate the early stages of precipitation and coarsening of carbide particles and nitride particles in several heat-treated steels, e.g. in isothermally transformed Fe-2 w% V − 0.2 w% C[229, 230] and in variously nitrided and aged Fe-Mo alloys[80, 231, 232]. During early stage transformation the carbide and nitride particles often are very small ($\lesssim 3$ nm in diameter) and are present at a high volume density ($\sim 10^{18}$ cm^{-3}). Such a finely dispersed microstructure is usually not amenable to a detailed analysis by TEM because of a resulting overlap of the strain fields of adjacent (often coherent) particles (Fig. 4.21 a). By using FIM, these problems are avoided and individual carbide particles (other than iron carbides[211]) or nitride particles are commonly imaged in *bright* contrast at tip temperatures above approximately 80 K (Fig. 4.21 b); imaging below 80 K often enhances the recognition of low index facets in the matrix but also results in a loss of the pronounced contrast between particles and matrix (Fig. 4.21 c).

Much *atom-probe work* in this area has focussed on the analysis of the partitioning of various steel constituents between matrix and precipitated particles (see also Chap. 4.5.3) and to the analysis of the composition of the carbide and nitride phases during

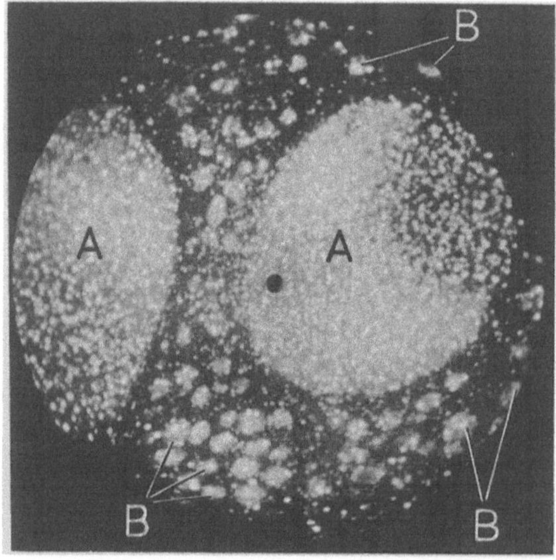

Fig. 4.20. Neon field ion image of a nickel based superalloy exhibiting two large (*A*) and many small (*B*) γ' precipitates. (Courtesy P. A. Beaven)

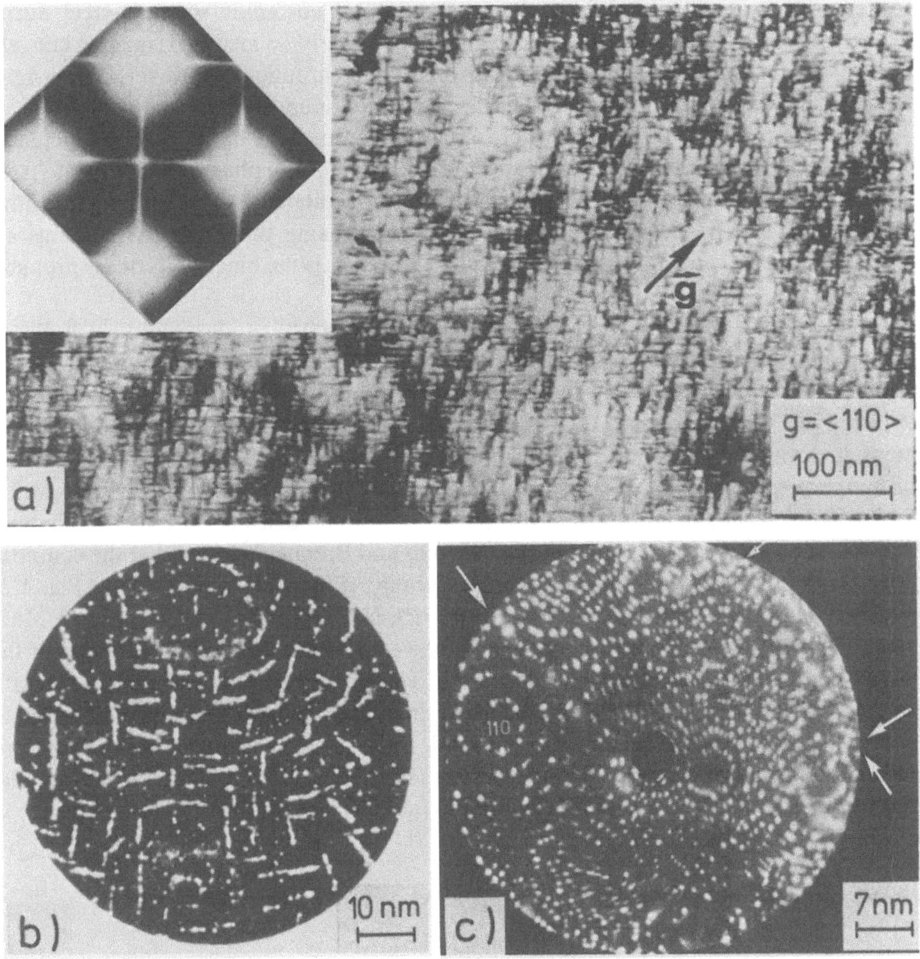

Fig. 4.21 a–c. TEM micrograph of nitrided Fe-3 at% Mo (**a**). Field ion images (neon, ∼ 9 kV) of the same alloy at about 100 K (**b**) and at about 50 K (**c**). In **b**) the matrix remains dark whereas in **c**) some low index poles can be recognized. *Arrows* in **c**) indicate the presence of nitride platelets[80]

aging [e.g. Ref. 211. 233]. It has been consistently reported[213, 234] that carbon field evaporates in various complexes having different ionization states such as C^{2+}, C^+ or C_2^{2+}, C_3^{2+}, C_2^+, and C_3^+. Since it is not possible to distinguish between C_2^{2+} and C^+ ions (both have a m/n-ratio of 12) it is impossible to obtain the *precise* stoichiometry of the carbides.

In a commercial drawn pearlite steel with additions of Mn and Si, Miller and Smith[213] determined the metal/carbon ratio of the carbide phase (cementite) to be within the range 3.07 to 2.74 ($\pm$ 0.18) depending on the assignment of the 12 a.m.u. peak. In a molybdenum steel, Turner and Papazian[234] analysed carbide platelets of about 10 nm in diameter and obtained a stoichiometric composition, which is very close to Mo_2C. Only recently, Adrén et al.[235] studied the compositional changes of MC-type carbides (M representing a metal atom) in a titanium stabilized austenitic stainless steel during isothermal aging at 700 and 750 °C. They report all *Ti carbides* analysed (ranging between 8 and 14 nm in size) to contain a significant amount of chromium. However, during aging at 700 °C the Cr content in the carbides decreased continuously from about 14 at% after 3 h to about 5 at% after 1130 h, whereas the carbon content in the carbides increased from 40 at% to that (50 at%) of the expected TiC equilibrium phase. Obviously at 700 °C nucleation of TiC particles in this particular steel is facilitated by first forming transition carbides rich in Cr and depleted in C; during coarsening the Cr and C contents are gradually adjusted and only after *extended* aging periods is the binary TiC phase probably obtained.

Nitriding of a Fe-3 at% Mo alloy between 450 and 600 °C produces a finely dispersed precipitated nitride phase consisting of thin coherent platelets on the $\{100\}_\alpha$ planes[80, 231, 232]. The platelets can be easily resolved in the FIM but not in the TEM as depicted in Fig. 4.21 b and 4.21 a, respectively. Isothermal aging of the nitrided specimens at elevated temperatures (performed at both 500 and 800 °C) generates a sequence of morphologically different phases (platelets, thick plates and spheres) as is shown in Fig. 4.22 a–f. Comparing the composition of 36 morphologically different precipitates that were analysed with the atom probe, Wagner and Brenner[80] found that the compositions could be subdivided into three distinct groups (Fig. 4.23). Inspection of Fig. 4.23 clearly reveals the occurrence of two nitrides *rich* in Fe (groups I and II); from a Mößbauer study of the same alloying system[236], however, it was concluded that none of the nitride phases contained more than 10 at% Fe. This discrepancy between the two investigations can only be explained on the basis that the assumptions, made in order to interprete the Mößbauer spectra, are not valid. Based upon structural and stoichiometrical considerations of conceivable nitride phases in ferrous alloys by Jack[237], Wagner and Brenner concluded from their atom-probe FIM studies that the precipitation sequence in the Fe-Mo-N system is as follows:

$$\alpha''\text{-}(Fe, Mo)_{16}N_2 \rightarrow \eta'\text{-}(Fe, Mo)_6N_2 \rightarrow \gamma\,(Fe, Mo)_2N \rightarrow \delta\text{-}(Fe, Mo)\,N$$

In this sequence, which progresses via three intermediate nitride phases (α'', η', and γ) before the (hexagonal) equilibrium phase (δ) is established, the Fe concentration of the precipitate continuously decreases, whereas both the N and the Mo content increase during the course of transformation.

The coarsening of the α''-nitride platelets at 600 °C was also investigated in this study and as can be seen (from Fig. 4.22 a–d the thickness of the platelets ($\lesssim 1$ nm) did not change markedly whereas their diameter increased from 6.6 nm directly after nitriding to 25.6 nm after 623 h. During the early coarsening stage the coarsening kinetics followed a $\bar{r}^3 \sim t_A$ time dependence indicating that coarsening is diffusion controlled; analysing the data in terms of the Wagner-Livshitz coarsening theory[224], modified for platelet shaped particles, yielded values for the interfacial energies of the edge and the habit plane of the platelets.

Although atom-probe analysis of single precipitates can be rather tedious and time consuming, it is so far the only microanalytical technique which provides comprehensive information about the change in both the morphology and chemistry of different phases formed during such a complex precipitation sequence like the one in nitrided Fe-Mo. On the other hand, one has to verify that the small specimen volume ($\sim 10^{-15}$ cm^3 as com-

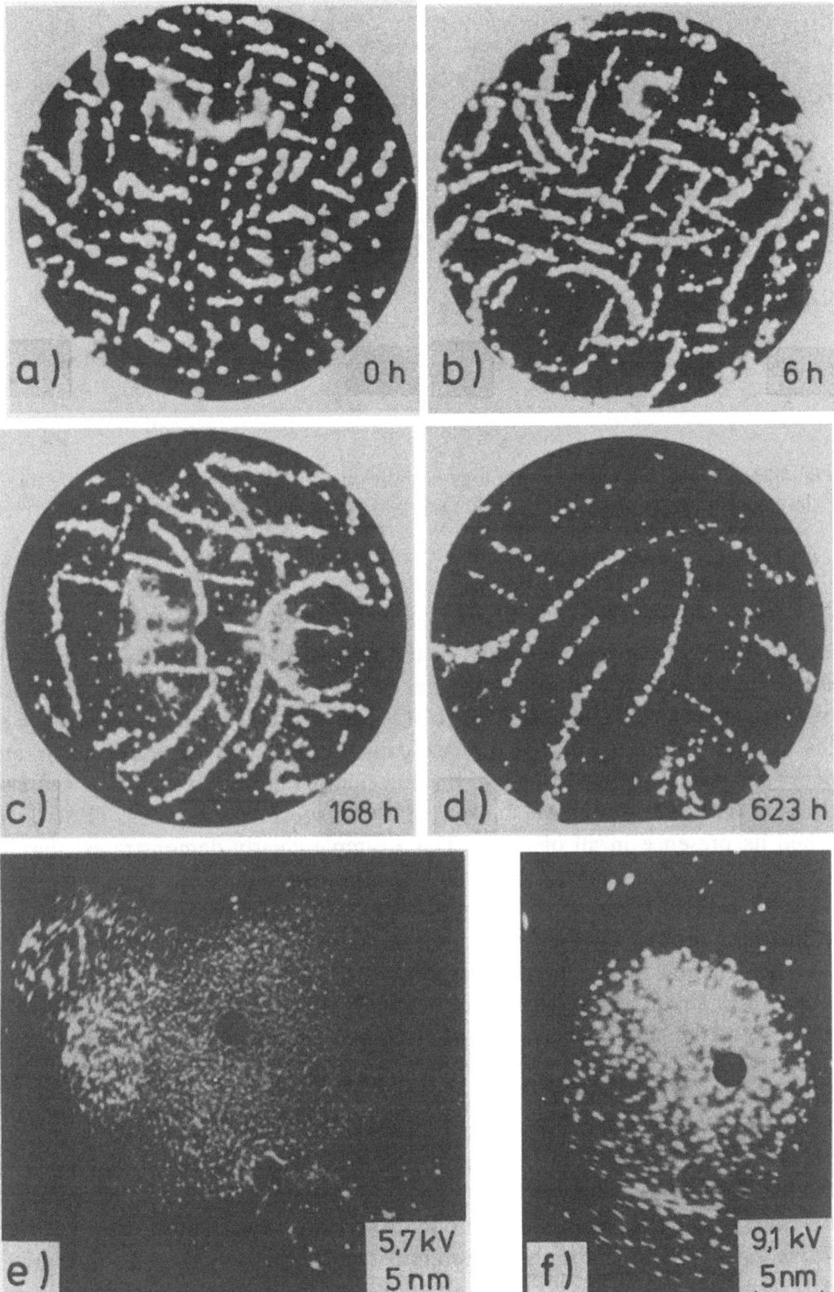

Fig. 4.22 a–d. Coarsening of nitride platelets during aging at 600 °C. **e)** Thick nitride plate after nitriding at 500 °C and aging for 1 h at 800 °C. **f)** Spherical precipitate after nitriding at 600 °C and aging for 1 h at 800 °C. With the chosen imaging conditions (neon, 80 K) the matrix remains dark[80]

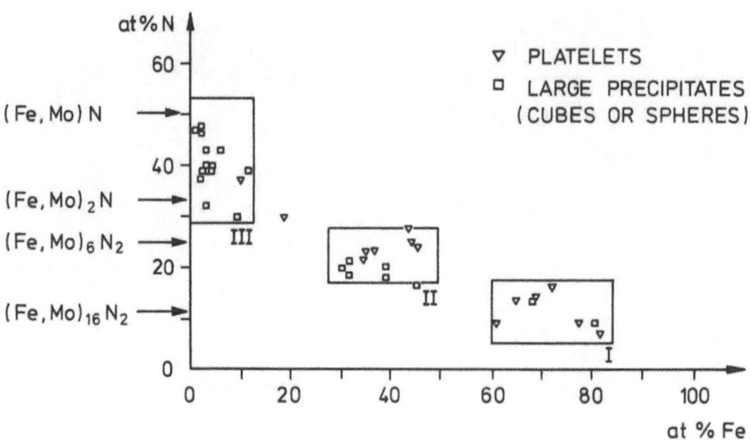

Fig. 4.23. Composition and morphology of individual precipitates in differently treated specimens determined by atom probe FIM. From each particle at least 200 ions were collected[80]

pared to $10^{-11}\,cm^3$ in the TEM), which is sampled during a typical atom-probe FIM experiment, is truly representative of the microstructure of the bulk material. For instance, in the later transformation stages of the Fe-Mo-N alloy, the α'' phase, composed of thin platelets with a number density of $\sim 5 \times 10^{16}\,cm^{-3}$, was simultaneously present in the material together with the δ phase consisting of large spherical precipitates with a density of only $\sim 10^{14}\,cm^{-3}$. Due to this low number density, a δ particle was observed in the FIM only *once* during the analysis of *twelve* specimens although TEM clearly established its presence in all of them. This example clearly demonstrates, that additional investigations of the same alloy employing a complementary technique (in general TEM) are often necessary, making the quantitative analysis of a precipitation reaction by means of the atom-probe FIM even more time consuming.

Since the FIM and/or the atom probe allow to determine the size of very fine precipitates, which are no longer discernible in the TEM, the technique is particularly suited for the investigation of the kinetic laws of *precipitate growth* and *precipitate coarsening* in the early stages of the phase transformation. According to Wagner[224] the *coarsening* of precipitates follows either a $\bar{r}^3 \propto t_A$ or a $\bar{r}^2 \propto t_A$ power law depending on whether bulk diffusion or whether the atom transfer across the interphase boundary, respectively, is the rate controlling mechanism. The $\bar{r}^3$ power law has been found frequently in various two-phase alloys by means of both TEM (e.g.[238]) and FIM (see above); the $\bar{r}^2$ kinetics, however, have been reported so far only from FIM investigations of the coarsening rate of fine V_4C_3 particles in the above mentioned vanadium steel[229] and of fine gold particles in a Fe-3.4 at% Au alloy[239]. In both studies the activation energy for the coarsening process as determined from an Arrhenius plot was shown to be significantly smaller than that expected for lattice diffusion of V or, respectively, of Au in ferrite.

Although the FIM results from both alloys suggested an interface-controlled coarsening rate of the fine precipitates ($\bar{r} \approx 1$ to 15 nm), the observation of a $r^2 \propto t_A$ power law during the early stages of a precipitation reaction can also be accounted for by the *diffusion-controlled growth* of very small particles into a still supersaturated solid solution (e.g.[240]). This has been shown by Goodman et al.[221] for the above mentioned Fe-1.4 at% Cu alloy in which the nucleated Cu-rich particles ($\bar{r}$ = 0.8 to 7 nm; Fig. 4.17) grew primarily by taking up copper atoms from the supersaturated ferrite matrix rather than by Ostwald ripening, i.e. coarsening. The diffusivity, as determined from the slope of a $\bar{r}^2$ vs. t_A plot, corresponded to that of Cu in ferrite; hence, the growth rate was concluded to be diffusion controlled.

4.6.3 Spinodal Decomposition

So far only a few FIM and atom-probe investigations dealing with the characterization of spinodal decomposition in supersaturated alloys have been reported. Many of these have been very preliminary in character and not very conclusive (e.g. in Co-Ti[74] and in Fe-Cr[241]) or even misleading as was discussed in the previous chapter with respect to Ni-Al.

CuNiFe is consered to be *the* "classical spinodal" system since there exists convincing evidence from comprehensive TEM and magnetic studies that these alloys transform in a continuous mode[242, 243]. In this alloy (containing 45.5 at% Ni and 22.5 at% Fe), Watts[244] attempted in one of the first combined FIM and atom-probe investigations to verify the sinusoidal picture of composition fluctuations as suggested by theory[218, 219] and to test the predicted pseudo-binary nature of the phase transformation. Stationary field-ion images of both the as-quenched and the aged specimens turned out to be rather poor and, therefore, did not provide significant information about the transformation products. Sequences of FIM images recorded on video tape during field evaporation, however, showed the copper-rich phase (appearing dark in the FIM patterns) to be highly interconnected during aging at 625 °C for less than 1 h; with further aging the interconnected phase is concluded to break up into discrete cubiodal and rod-shaped precipitates, at this stage also discernible in the TEM[243].

Attempts to measure composition profiles in CuNiFe with the atom-probe failed completely because a drastic reduction in the detection efficiency for copper ions was found; most atom-probe analyses yielded a copper concentration of less than 10 at% by comparison to the nominal 32 at%. The variation of several experimental parameters, such as tip temperatures, pulse height, evaporation rate, and also the partial pressure of hydrogen, did not improve the poor detection efficiency and, hence, CuNiFe seems not to be amenable to atom-probe analyses. The reason for this behaviour is still not well understood. Watts found some evidence that the formation of copper-ion complexes, each ion of which is singly charged, might be to some extent responsible for the reduced detection efficiency; in this case the complex would be recorded as only one single Cu^+ ion (see Chap. 3.1). However, if this were the true explanation the question remains open as to why this effect is not observed in other copper based alloys, e.g. in Cu-Ti.

The only successful and comprehensive atom-probe FIM studies of spinodal decomposition have been performed on Ni-Ti[151] and on Cu-Ti alloys[89] and we shall therefore concentrate in the following on the results and discussion of these.

Isothermal aging of Ni-Ti alloys at 600 °C (with Ti contents ranging between about 10 and 14 at%) results in the formation of sidebands and a quasi-periodical arrangement of cuboidal (metastable) γ'-Ni$_3$Ti precipitates as observed in X-ray diffraction patterns[245] and by means of TEM[246]. Based on these observations it was concluded that phase separation in Ni-Ti occurs via a continuous transformation, i.e. a spinodal reaction. In order to verify this conclusion, Sinclair et al.[150, 151] investigated the decomposition of a supersaturated Ni-14 at% Ti alloy by employing jointly FIM and X-ray diffraction techniques. It was shown[150] that *ordered γ'* particles appear in bright contrast if imaged in neon *with* a trace of hydrogen. Furthermore, contrast analysis of the ordered γ' phase revealed that only nickel atoms are imaged under these conditions, i.e. successive field evaporation of superlattice planes (e.g. {200} or {220} in the LI_2 crystal structure of γ') results in alternating brightly imaged rings (from planes containing only Ni sites) and dim rings (from planes containing atomic sites in the ratio 2 Ni : 2 Ti). The latter "mixed" planes are much less stable to field evaporation than the pure nickel planes. Therefore, as was discussed in Chap. 4.4 the *plane stability ratio R* (see Eq. 4.3) is smaller for the ordered regions than it is for the disordered matrix (R = 1) and its measurement provides information about the changes in local order and composition of the ordered regions during an aging sequence[151].

During aging at 600 °C both the shape and the distribution of the ordered regions became more regular approaching a modulated microstructure in the later stages. Studies of the atomic arrangement within the ordered regions revealed that during the early aging stages many Ni atoms occupied sites of the Ti sublattice whereas *only few* Ti atoms were observed on the Ni sublattice. Counting the number of misplaced atoms yielded the Ti concentration in the ordered regions to be 18 ± 4 at% after aging for only 15 min and to be 23 ± 4 at% after 16 h; these figures must be compared with 18 ± 5 at% as determined with the atom probe for the brightly imaged regions in *as-quenched* specimens. Figure 4.24 shows the variation of the plane stability ratio R with distance during aging. It

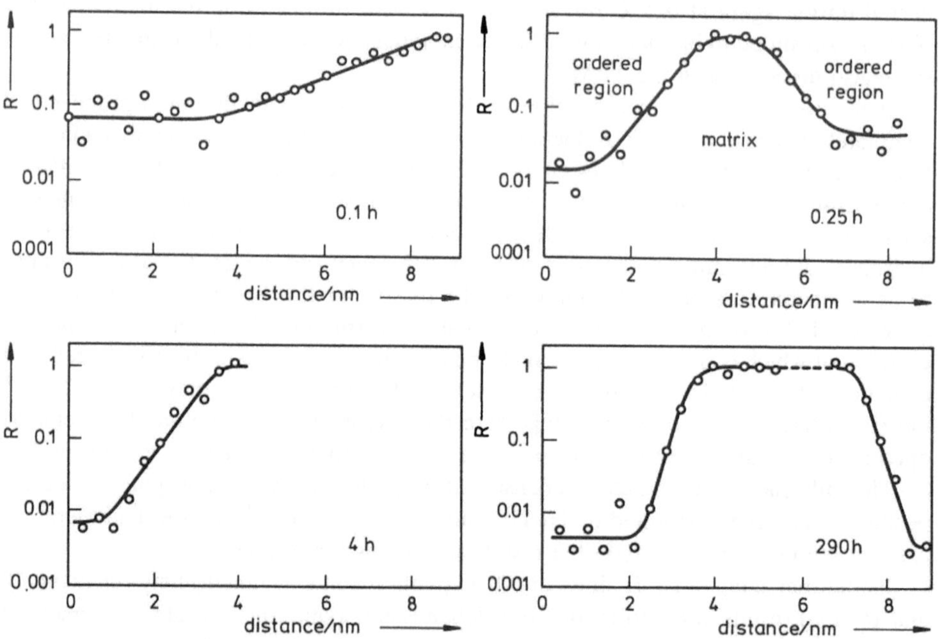

Fig. 4.24. The variation of plane stability ratio *R* with distance along a ⟨100⟩ direction for Ni-14 at% Ti aged at 600 °C for the given periods (After R. Sinclair et al.[151])

can be seen that R varies smoothly between the ordered regions ($R_{ord} \ll 1$) and the matrix ($R_m = 1$); furthermore, the plateau value of R_{ord} in the ordered regions *decreases* continuously with aging time. This has been taken as direct evidence that the degree of order continuously increases. Since this can only happen through the replacement of the originally misplaced nickel atoms on the Ti sublattice, this result also indicated that the Ti content in the ordered regions gradually increases with aging time, and, hence, that phase separation takes place via the spinodal mode.

Figure 4.24 also shows a considerable reduction of the interphase boundary width after aging for 20 h; however, the boundary is not atomically sharp. Both the size of the ordered regions and the modulation wavelength λ_{FIM} (corresponding to the centre-to-centre spacing) along $\langle 100 \rangle$ were determined from persistence length measurements (see Chap. 2.7.1) and also directly from the FIM micrographs. Whereas the two differently determined values λ_{FIM} agreed quite well, they turned out to be almost a factor of two smaller than the figures ($\lambda_{X\text{-ray}}$) obtained from X-ray sideband analyses.

This discrepancy was explained by stating that the two techniques measure different quantities; i.e. one the modulation of the composition and, hence, that of the lattice parameter (X-ray) and one the modulation of the local order (FIM). If this *were* the true explanation it would mean that ordering *precedes* compositional segregation. In this case there might indeed exist some regions whose degree of order is still sufficient to yield contrast in the FIM but whose Ti content is too small to produce a sufficient difference in lattice spacing and, thus, in scattering intensity. Therefore, sideband analyses would lead to a larger modulation wavelength. The crucial conclusion that ordering precedes clustering was drawn more explicitly in a subsequent paper by Watts and Ralph[247] based on both the above results and on further atom-probe analyses of aged Ni-12 at% Ti, but has recently been criticized by Laughlin et al.[248]. It therefore seems that further experiments are necessary in order to solve the discrepancy between λ_{FIM} and $\lambda_{X\text{-ray}}$.

Although the significance of the quantitative measurements is debatable, the qualitative results obtained from these FIM studies support the earlier conclusions that Ni-Ti decomposes spinodally.

Some controversy exists concerning the mode of phase transformation in Cu-Ti alloys with Ti contents ranging between 2 and about 5.5 at% during isothermal aging between 300 and 500 °C. Recent microstructural investigations by means of TEM[249] have suggested a spinodal reaction resulting finally in a modulated structure consisting of ordered metastable Cu_4Ti precipitates. In contrast, results obtained from X-ray scattering experiments[250] suggested the formation of three-phase zone complexes with a diffuse periodicity in the early decomposition stages; furthermore, from this study it was concluded that particles having the stoichiometric composition Cu_4Ti will *not* form as the final transformation product during isothermal aging below 400 °C.

This controversy over the mechanism of the phase transformation prompted Biehl and Wagner[89, 251] to analyse the mechanism and the kinetics of early stage decomposition in Cu-2.7 at% Ti at 350 °C by means of the atom-probe FIM. Since the field-ion images did not provide much information about the microstructural states (see Chap. 2.6.3 and Fig. 2.14), the investigation was primarily based on autocorrelation (Chap. 3.2; Eq. 3.2) and Fourier analyses of composition profiles extending typically over more than 400 successively field evaporated {111} planes; this corresponds to a depth probing of more than 80 nm which – as will be shown later on – is always more than ten times the modulation wavelength in this system.

Autocorrelation analyses of homogenized specimens yielded the correlation length k_0 to be zero, indicating that no decomposition occurred during the quench. During aging k_0 and, hence, the diameters of the Ti-rich regions increased in proportion to $t_A^{1/3}$ (Fig. 4.25) as predicted by Langer[220] for spinodally decomposing alloys. After aging for 50 min the composition profiles (e.g. Fig. 3.2) clearly revealed the existence of discrete β'-Cu_4Ti precipitates, thus confirming the earlier TEM results of Laughlin and Cahn[249] and proving the X-ray analyses of Tsujimoto[250] to be incorrect. The correlation length k_1 (Fig. 3.3 and Chap. 3.2), corresponding to the interparticle distance λ_{111} along a $\langle 111 \rangle$ direction, was found to coarsen immediately following a $\lambda_{111}^4 \propto t_A$ power law (Fig. 4.26) in agreement with X-ray sideband analyses of Miyazaki[252]. The appearance of a second peak in the autocorrelograms (Fig. 3.3) at $k_2 = 2\,k_1$ indicated the Cu_4Ti particles to be quasi-periodically arranged on a sort of macro lattice of lattice parameter λ; this is directly corroborated by the FIM image (Fig. 2.14 b) in which the Cu_4Ti particles (dark contrast) are quasi-periodically aligned along the three $\langle 100 \rangle$ directions.

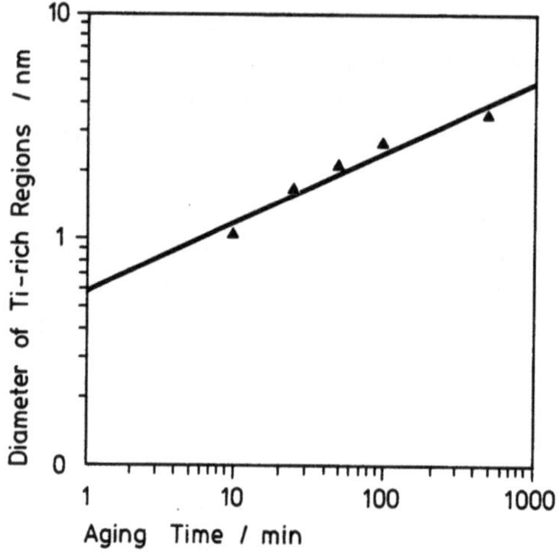

Fig. 4.25. Growth of the Ti-rich regions with time during aging at 350 °C[89]

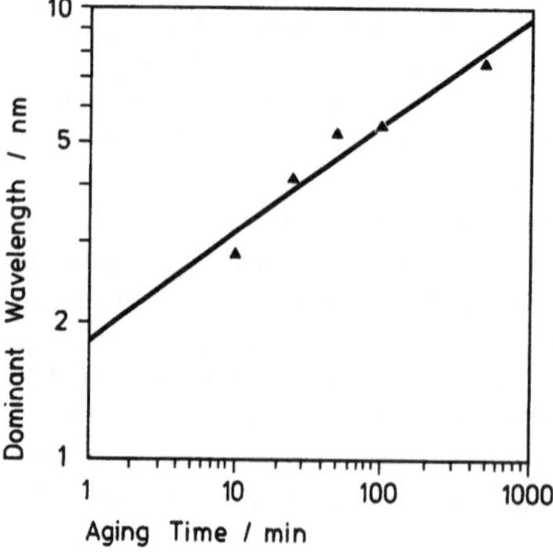

Fig. 4.26. Growth of the average modulation wavelength λ with aging time in Cu-2.7 at% Ti aged at 350 °C. λ was determined from both autocorrelation analysis and Fourier transforms of composition profiles[89]

Although these microstructural observations already strongly suggested a spinodal reaction to be the transformation mode in Cu-2.7 at% Ti, it was still necessary to follow the evolution of the composition of the Ti-rich regions with aging time. This turned out to be much more complicated than the measurement of their sizes and of λ, as the following more general discussion will demonstrate.

According to theories of spinodal decomposition[218, 219], the incipient modulated structure observed in this Cu-Ti alloy is predicted to evolve from the superposition of three sinusoidal composition waves of equal amplitude $A_0/3$ and wavelength λ extending along the three $\langle 100 \rangle$ directions x_i (i = 1, 2, 3); the resulting structure of the macro-lattice is then of the CsCl type with the centres of the Ti-enriched regions and the Ti-depleted regions on Cs sites and Cl sites, respectively. In visualizing such a periodical three-dimensional structure, it becomes clear that the *amplitude* of composition profiles as recorded with the atom probe must depend on the choice of both the effective diameter d_{ap} of the probe hole with respect to λ (see also Ref. 253), and on the position of the analysed cylindrical volume within the macro lattice. (It should, however, be emphasized that the diameters of the Ti-rich regions, i.e. k_0, and the modulation wavelength, i.e. $\lambda = k_1/\sqrt{3}$, as determined from autocorrelation analyses are independent of the choice of this set of parameters.) In order to investigate this dependence more quantitatively and, hence, to find the optimum experimental parameters, Biehl and Wagner[89] simulated atom-probe analyses of modulated structures on a computer. For this purpose three-dimensional composition variations in a fcc solid solution of average concentration c_0 of solute atoms were approximated by

$$c_0(x_1, x_2, x_3) - c_0 = \frac{A_0}{3} \sum_{i=1}^{3} \sin \frac{2\pi}{\lambda} x_i \qquad (4.4)$$

In an analogous way to real atom-probe experiments, composition profiles from a cylindrical volume whose axis was parallel to [111] were generated by computer; the diameter of the cylinder and its position within this idealized modulated structure could be varied. The amplitude c_{max} as obtained from the simulated composition profile was then compared with the true amplitude $A_0^* \equiv A_0 + c_0$, corresponding to the composition of the solute-rich regions.

In the first simulation experiment the wavelength λ of the modulated structure (Eq. 4.4) was varied between 3 and 12 nm, with the axis of the cylinder of analysis always being chosen to run through the centres of the solute-rich regions. As a result, in Fig. 4.27 the normalized composition amplitude c_{max}/A_0^* from corresponding composition profiles is plotted as a function of λ for two different sizes d_{ap} of the probe hole. The simulations reveal immediately that for all expected wavelengths $\lambda \leq 12$ nm the "measured" amplitude c_{max} of the solute-rich regions remains below the true amplitude A_0^*. The discrepancy turns out to be smaller, the larger the wavelength and the smaller d_{ap} are chosen. However, as was discussed in Chapters 3.2 and 3.3, the size of the probe hole can not be chosen smaller than ~ 1.5 nm since statistics require a sufficient number of atoms to be collected per field evaporated plane. Thus one has to accept that for $\lambda = 3$ nm the measured amplitude c_{max} deviates by at least 19% from A_0^*. This deviation can be significantly larger if the cylinder axis no longer runs through the centres of the solute-rich regions. This was demonstrated in a second simulation experiment in which a cylinder of

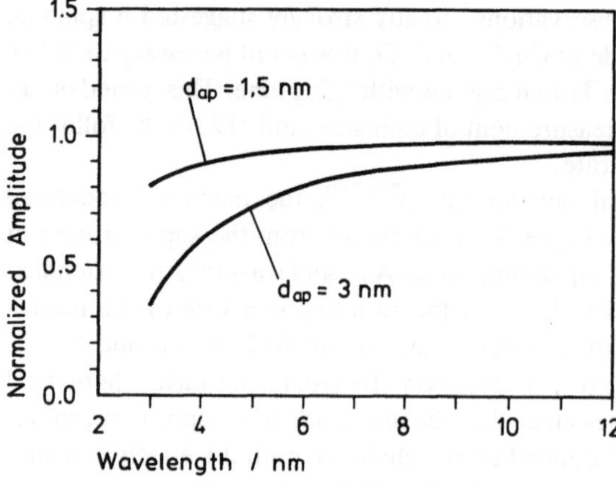

Fig. 4.27. The variation of the composition amplitude (normalized with respect to A_0) with modulation wavelength for two different sizes of the probe-hole as obtained from computer simulations. The cylinder of analysis lies parallel to $\langle 111 \rangle$ and cuts through the centres of the solute rich regions. (After K.-E. Biehl[251])

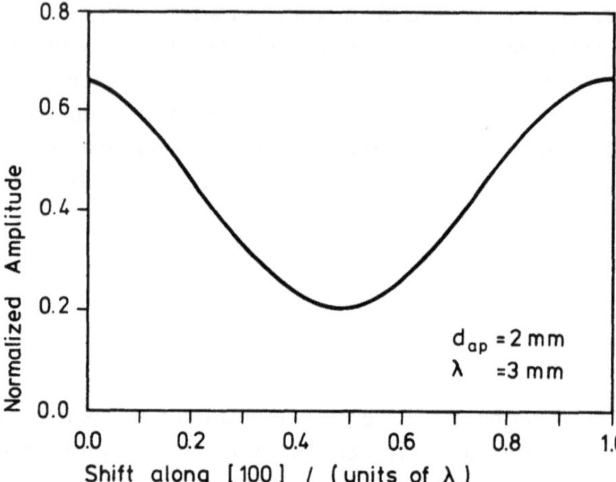

Fig. 4.28. The variation of the normalized composition amplitude with position of the analysed cylindrical volume within the macrolattice of lattice parameter $\lambda = 3$ nm. (After K.-E. Biehl[251])

fixed diameter ($d_{ap} = 2$ nm) was shifted along a $\langle 100 \rangle$ direction through a modulated structure of constant wavelength ($\lambda = 3$ nm) (Fig. 4.28). If the cylinder is placed at 0.5 λ, i.e. its axis runs through the centres of the solute-depleted regions, c_{max} is measured as only 20% of the true amplitude A_0^*.

Using the results depicted in Fig. 4.27 and 4.28 it is in principle possible to derive the amplitude of a sinusoidal fluctuation from c_{max} of a measured (smoothed) composition profile, provided the position of the cylinder of analysis can be determined within the macro-lattice. This is not a straight-forward procedure and can only be achieved empirically by taking several composition profiles from either different specimens (with the cylinder axes always running parallel to the same crystallographic direction and d_{ap} being kept constant), or from the same specimen with different positions of the probe-hole. (In the latter case the probe-hole must be shifted in fractions of λ, the value of which can be obtained from autocorrelation analysis of the first profile.) The composition profile which reveals the maximum values of c_{max} is then the one obtained with the cylinder of analysis cutting through the centres of solute-rich regions. Although this approach does promise rather accurate values of A_0, it is rather tedious and has not yet been applied comprehensively to the analysis of the composition amplitude of spinodal alloys.

For most spinodally decomposing alloys c_0 is asymmetrical with respect to the (coherent) phase boundaries. For these alloys the sinusoidal picture is therefore only appropriate during the early stages of spinodal decomposition. In the later stages of the reaction, the amplitude of the composition waves are no longer symmetrical with respect to c_0 and the volume fractions, i.e. the spatial extensions, of the two evolving phases are in general no longer equal; furthermore the modulated structure will contain numerous imperfections in its periodicity[254] which have been suggested to be necessary for coarsening[219]. In such an imperfect modulated structure the cylinder of analysis will always cut through one or several solute-rich regions. The maximum concentration of these incipient particles found in a smoothed composition profile of sufficient depth (e.g. $\gtrsim 10 \cdot \lambda$), may then be considered to be a reasonable measure of their composition. As in the case of discrete precipitates, this procedure for the evaluation of the composition amplitude of solute-rich regions (size d) does require the ratio d_{ap}/d to be smaller than unity.

According to the autocorrelation analyses (cf. Fig. 4.25) this condition is not fulfilled by Cu-2.7 at% Ti alloys aged for 10 and 25 min at 350 °C. Under these circumstances the apparent composition of the Ti-rich regions must be smaller than the 20 at% Ti required for Cu_4Ti precipitates, and is expected to increase with decreasing d_{ap}/d during coarsening, regardless of whether the alloy already contains Cu_4Ti precipitates or of whether decomposition has occurred spinodally. Thus, for $d_{ap}/d > 1$ it is not possible to distinguish a spinodal reaction from a classical precipitation reaction by following the time evolution of the frequency distribution of Ti-concentrations in smoothed composition profiles. In this case the only feasible method of deconvoluting the composition of the Ti-rich regions is by simulating atom-probe experiments of realistic precipitation microstructures on a computer and comparing the resulting composition profiles with the experimental ones. Using this approach, Biehl simulated atom-probe analyses ($d_{ap} \simeq 2.5$ nm) with a precipitation structure already containing Cu_4Ti-particles; for this structure he observed no correspondence between the simulated and the experimental composition profile. In contrast, for the specimen aged for 10 min he obtained rather good agreement between the two profiles by assuming a sinusoidal spinodal structure of wavelength $\lambda \simeq 2.6$ nm, as derived from autocorrelation analyses (Fig. 4.26), and of amplitude $A_0 \simeq 5.1$ at% corresponding to a Ti-concentration of about 7.8 at% within the Ti-enriched regions (Fig. 4.29). The sinusoidal picture as used for the simulations appears to be appropriate for the specimens aged for 10 min, where the amplitude ($A_0/3$) of a one-dimensional wave is evaluated as ~ 1.7 at% indicating that the decomposition is still symmetrical with respect to $c_0 = 2.7$ at%. After aging for 25 min $A_0/3$ is evaluated as 3.7 at% and a sinusoidal wave is no longer a good approximation for the composition modulations. Therefore, the reported value of $A_0^* = 13.8$ at% Ti (Fig. 4.29) must be regarded with some reservations. The fact, however, that after aging for 10 min the Ti-concentration of the Ti-enriched regions is found to be much less than 20 at% Ti provides convincing evidence that spinodal decomposition is the transformation mode in a supersaturated Cu-2.7 at% Ti alloy. The spinodal reaction is definitely terminated after about 50 min at which time metastable Cu_4Ti-particles are directly discernible (e.g. Fig. 3.2) in the composition profile.

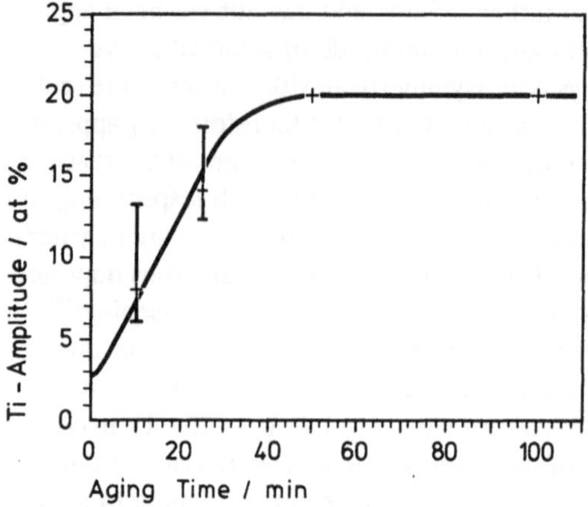

Fig. 4.29. Variation of the Ti content of Ti-rich regions with time during aging at 350 °C. (After K.-E. Biehl[251])

4.7 Semiconductors

From several studies of the imaging characteristics of whiskers of Si[255], Ge[256], and also of GaAs[257] and GaP[258, 259] in the FIM it has been consistently reported that the order in the field-ion images of these semiconductors is rather poor by comparison to FIM images of metals. The image quality is somewhat improved by adding some hydrogen to the imaging gas, but irrespective of the choice of the latter (Ar, Ne or He) the ring structure especially around the {100} regions is barely developed (Fig. 4.30); furthermore with additions of hydrogen no stable field ion images could be obtained.

Due to the high electrical resistance of high purity semiconductors the field penetration depth λ for these materials is much larger than that for metals. As a consequence, the field evaporated ions suffer an energy loss which may amount to some hundreds of electron-volts[257]. In addition, the field evaporation pulse is significantly deformed if applied to a tip of large resistance; this effect leads to further energy deficits of the field-evaporated ions. The degradation of the mass resolution associated with these energy deficits (see Chap. 3.3) precludes the analysis of *high-purity* semiconductors with the conventional straight ToF atom probe. The problems can be circumvented to a large extent by employing the energy-focusing atom probe (Chap. 3.4). This was demonstrated by Melmed et al.[261] for silicon whiskers with different resistivities. Although the energy deficits of the field evaporated Si^+ ions reached up to a few hundred eV, meaningful mass spectra could be obtained from all specimens provided the pulse duration was chosen to be sufficiently long. For lower resistivity silicon ($\lesssim 600$ ohm-cm) a pulse width of 17 ns was found to be sufficient, whereas for the highest resistivity specimens (2.5×10^4 ohm-cm) the pulse duration had to exceed 300 ns.

Yamamoto and Seidman[259] dissected a tip of low resistivity GaP (4×10^{-3} ohm-cm) along a $\langle 111 \rangle$ direction by pulsed field evaporation and verified with the straight ToF atom probe that alternating {111} planes consist of either Ga or P atoms as expected from the underlying zinc-blende structure. Sometimes, however, a Ga or a P atom was found to be misplaced on a P plane or a Ga plane, respectively. Keeping both the pulse fraction ($V_P/V_{DC} = 0.15$) and the pulse frequency (60 Hz) constant, they observed the

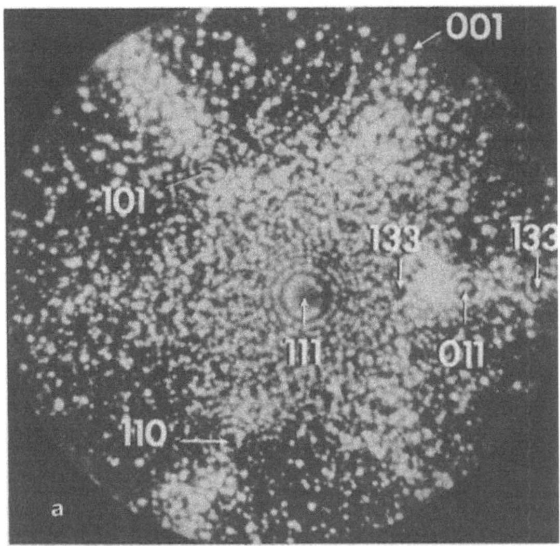

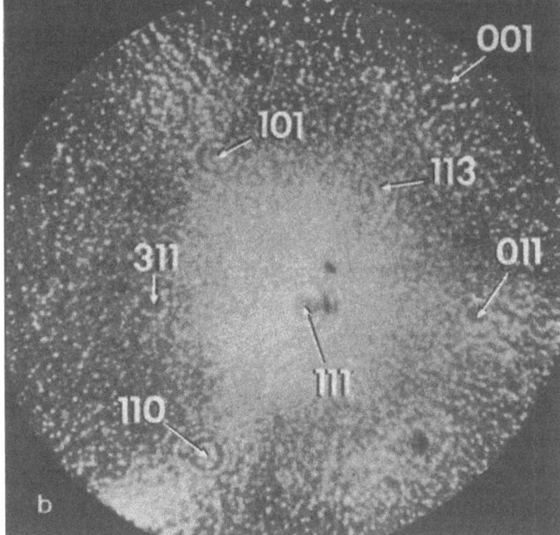

Fig. 4.30 a, b. Field-ion micrographs of silicon. The tip temperature was ~ 25 K. Imaged in a mixture of hydrogen and neon at 12.5 kV (**a**), and in a mixture of hydrogen and helium at 17.5 kV (**b**). (Courtesy A. J. Melmed[255])

overall composition to be strongly dependent on the choice of the counting rate (which is related to the field evaporation rate); this was expressed in terms of the number of detected ions (N_i) for a given number of pulses (N_p). Only if N_i/N_p was chosen to be fairly small (~ 0.01) did the experimentally determined composition approach the expected stoichiometry; for larger values of N_i/N_p a drastic deficiency of P atoms was found.

The results presented above indicate that the problems associated with the analysis of semiconductors in the FIM and especially in the atom probe seem to be inherently more complex than with metals and alloys. At present it seems questionable whether the quality of field-ion images of semiconductors can ever be improved to such an extent that FIM analyses of point defects (e.g. self-interstitials or divacancies) or other small sized defects (e.g. B-type swirls) may become feasible.

With respect to the energy deficits of field-evaporated ions, which cause most problems in analysing semiconductors with the high-voltage pulse operated ToF atom probe, some promising progress has been made recently by Kellogg and Tsong[263]. Based upon earlier observations by Melmed and co-workers[255, 262], who reported an enhanced FIM-image brightness as well as an increased field-evaporation rate when high-purity semiconductors (Si, Ge, GaAs) are illuminated with white light, Kellogg and Tsong field evaporated a silicon specimen by illuminating the tip, held at a given DC voltage, with laser pulses of 6 nsec duration. Analyses of the field-evaporated silicon ions in a *straight* ToF atom probe yielded a mass resolution which is comparable to that obtained for metals conventionally field evaporated by applying high-voltage pulses to the FIM specimen. Although it is not yet clear whether the pulsed laser-stimulated field evaporation of *semiconductors* is dominated by either photon excitation effects or thermal effects, these preliminary results are rather encouraging for future studies of high-resistivity materials with a laser pulse-operated straight ToF atom probe.

In this context, it should be pointed out that a laser pulse applied to a metallic specimen instantaneously increases the tip temperature and therefore stimulates thermally assisted field evaporation (Chap. 2.4, Eq. 2.4). Therefore, metals and alloys can also be analysed with a laser-pulsed atom probe, which, in fact, might render some benefit for the following reasons: i) thermally assisted field-desorbed ions are always exposed to a constant electric field and, thus, unlike ions field evaporated by a high-voltage pulse, do not suffer any energy deficits (Chap. 3.1). Since the energy deficits of the ions limit the mass resolution of the straight ToF atom probe, its mass resolution ought to be improved by using a laser pulse. ii) Employing a laser pulse allows field evaporation at higher temperatures where the field-evaporation process is known to occur more smoothly whilst still enabling the field-ion specimen to be imaged at low temperatures where optimum resolution is achieved. Furthermore, since the electric field and, hence, the mechanical stress experienced by the specimen is not increased during the laser-stimulated field-evaporation event, the fracture susceptibility of the field-ion tip is considerably lowered.

Due to the heating of the tip by the laser pulse, surface diffusion may occur resulting in a redistribution of surface atoms prior to field evaporation. The effect, which has been observed in tungsten[263], might lead to incorrect results, for instance, in the analysis of thermally unstable solid solutions. However, the extent of this effect, and also that of the before mentioned ones which are considered to be beneficial, has still to be investigated systematically.

4.8 Metallic Glasses

Until now, both FIM and atom-probe studies of amorphous metals have been confined to metallic glasses consisting of about 80% transition metal atoms (T) and 20% metalloid atoms (M), such as $Pd_{80}Si_{20}$[264], $Fe_{80}B_{20}$[265, 266], and $(Fe, Ni)_{80}B_{20}$[90, 264]. Neon field-ion images of these $T_{80}M_{20}$ glasses (imaging gas pressure of about $1 \cdot 10^{-6}$ mbar, tip temperature below about 85 K) do not reveal the *regular* arrangement of image spots typically observed for crystalline metals and dilute solid solutions (Fig. 4.31). By comparison to crystalline materials imaged under similar conditions, the image spots of individual atoms

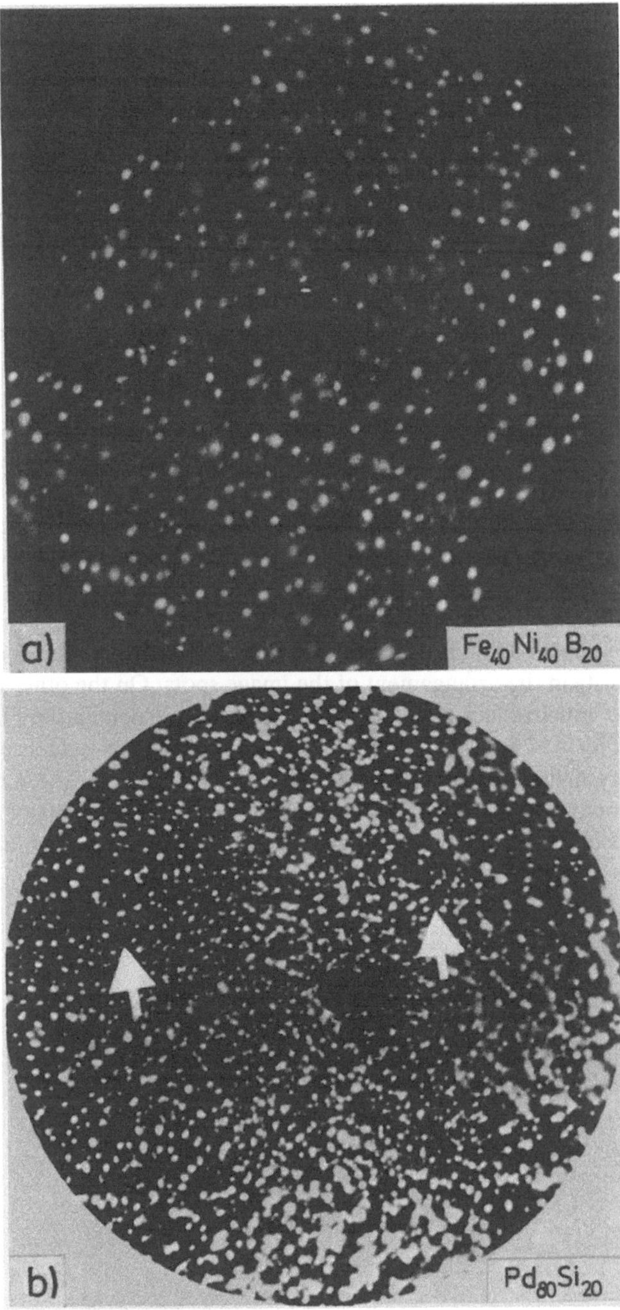

Fig. 4.31 a. Field-ion image of $Fe_{40}Ni_{40}B_{20}$; tip temperature ~ 90 K; imaged in neon at ~ 10 kV (Courtesy J. Piller). **b)** $Pd_{80}Si_{20}$; tip temperature ~ 50 K; imaged in neon at ~ 9 kV. *Arrows* point towards regions which might suggest the appearance of a ring structure

appear in stronger contrast and their diameters are found to be markedly increased; furthermore, the number of bright image spots is significantly decreased. This latter effect has been attributed to the fact that only one atomic species, i.e., either T or M atoms, is imaged. Inal et al.[265] and Jacobaeus et al.[266] assumed in their FIM studies of $Fe_{80}B_{20}$ and $Fe_{78}Mo_2B_{20}$ that only the transition metal atoms were discernible. This assumption has been proven wrong recently by Piller[90], who showed for $Fe_{40}Ni_{40}B_{20}$, employing the atom probe, that in fact about 80% of the brightly imaged atoms are boron atoms. Considering this result, one may conclude that due to a strong T-B binding the boron atoms are preferentially retained at the tip suface of the $T_{80}B_{20}$ metallic glass and therefore appear in enhanced contrast (see Chap. 2.6.1 and Fig. 2.4). The stronger resistance of boron (and also of silicon in $Pd_{80}Si_{20}$) against field evaporation manifests itself also in the fact, that the boron concentration as determined with the atom probe only corresponds to the nominal one if the ratio V_p/V_{DC} is chosen to be larger than 0.15; for smaller ratios the boron concentrations turns out to be significantly higher. These rather complex imaging characteristics already explain by themselves the absence of concentric ring structures in the field-ion images of $T_{80}M_{20}$ metallic glasses. Therefore the apparent non-existence of *regular* ring structures can not be taken as sufficient proof for the amorphous structure of the metallic glasses, as has been done in the past[265, 266]; in fact, as discussed earlier (Chap. 2.6.1), field-ion images of *crystalline*, concentrated, disordered, solid solutions (Fig. 2.12a) or those of steels, the matrix of which is distorted by small additions of interstitially dissolved carbon atoms, also do not contain any regularity in the arrangement of the image spots. On the other hand it is sometimes possible to visualize locally a distorted ring of image spots in the FIM patterns of a $T_{80}M_{20}$ glass (Fig. 4.31b). This, in turn, can not be taken as evidence for the existence of microcrystallites as was done by Inal et al.[265], especially for thermally treated $Fe_{80}B_{20}$; a computer-generated *random* distribution of image spots of about the same areal density as observed in $Fe_{40}Ni_{40}B_{20}$ also allows local occurrences of distorted ring structures to be imagined (Fig. 4.32).

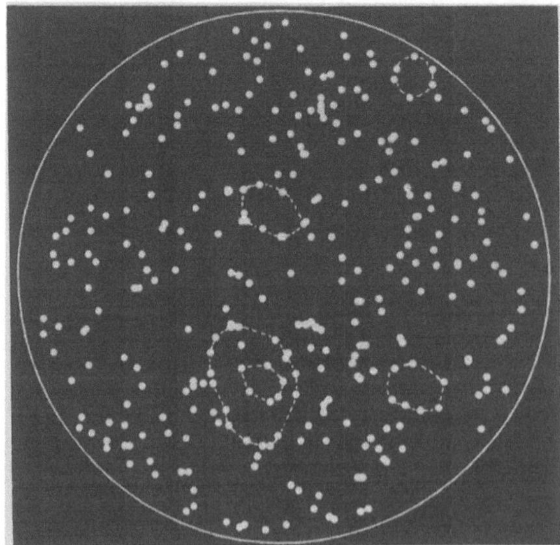

Fig. 4.32. A computer-generated random distribution of image spots with an areal density corresponding to that of brightly imaged boron atoms in $Fe_{40}Ni_{40}B_{20}$. Locally the occurrence of ring structures (*dashed*) can be visualized. (J. Piller[90])

Taking into account the complexity of the image formation mechanism and the rather complex projection of the curved tip surface onto the screen of the FIM, it appears impossible to obtain from field-ion micrographs any information about the correlation of atomic positions and, hence, about the amorphous structure of a $T_{80}M_{20}$ glass. (Although not yet explored it seems most likely that this conclusion holds true also for the class of TT metallic glasses containing no metalloid component.) The pair and the triplet correlation functions of the positions of transition metal atoms as empirically determined from field-ion micrographs of glassy $Fe_{80}B_{20}$ and $Fe_{78}Mo_2B_{20}$ by Jacobaeus et al.[266] must be considered to be not really representative for the amorphous structure of these alloys, if for no other reasons than because of erroneously assuming the iron and molybdenum atoms to be the imaging species.

For the same reasons outlined in Chap. 3.4 it is also not possible to investigate short-range ordering in the metallic glasses utilizing either the atom probe or the imaging atom probe. By analogy with crystalline alloys it is, however, possible to analyse short-ranged (cluster or precipitates of less than 1 nm diameter) as well as long-ranged (up to several tens of nm) compositional fluctuations in metallic glasses, e.g. resulting from a decomposition of the amorphous alloy during aging below the crystallization temperature. Since the lattice-plane counting technique (Chap. 2.7.1) is not applicable, the scaling of the probing depth in metallic glasses is not as straightforward and as accurately performed as in most crystalline materials, but rather has to be carried out by assigning a certain number $\overline{N}$ of collected ions to a certain probing depth. By reference to field evaporation of most densely packed planes (hkl) in crystalline metals at fixed experimental parameters (e.g. voltage and tip-to-screen distance), $\overline{N}$ is taken as the known average number of ions collected from each of these planes, and, thus, corresponds to their interplanar distance d_{hkl}.

In his atom-probe study (d_{ap} = 1.5 to 3 nm, $\overline{N}$ = 50 ions corresponding to 0.2 nm $\pm$ 0.02, V_{DC} = 8 to 11 kV) of composition fluctuations in $Fe_{40}Ni_{40}B_{20}$ in the as-quenched state and after isothermal and isochronous annealing below 370 °C, Piller[90] found stoichiometric $(Fe, Ni)_3B$ precipitates already in the as-quenched glassy alloy. Their radii, originally ranging around 1.5 nm, coarsened during isothermal annealing at 350 °C for 300 min to about 4 nm (Fig. 4.33 a); during this period the number density of the boron-rich clusters remained about constant at $(2 \pm 1) \cdot 10^{18} cm^{-3}$ (Fig. 4.33 b); it did not change noticeably during isochronous annealing for 5 h between 250 and 350 °C, but increased by a factor of about 10 during aging at 370 °C. Furthermore, it was found that the amorphous $(Fe, Ni)_3B$ clusters are embedded into amorphous regions extending up to 100 nm, whose average boron content deviated by as much as ± 3 at% from the nominal boron concentration (20.1 at%) (Fig. 4.34).

In regions of higher boron content (between 20 and ~ 24 at %) the mean cluster size was determined to be significantly larger than in the low-boron regions; in regions of only 17 at % boron essentially no decomposition took place (Fig. 4.34).

Due to the rather small difference in the boron concentration between amorphous matrix and $(Fe, Ni)_3B$ clusters, and their small sizes (allowing only a limited number of ions to be collected) the precipitates disappeared within the statistical fluctuations superposed on the boron concentration profiles and, therefore, could not be detected directly. The existence of the boron-rich clusters became, however, evident in the autocorrelograms of the concentration profiles from which their mean diameter could be determined (Chap. 3.2). Further parameters, which characterize the state of decomposition apart

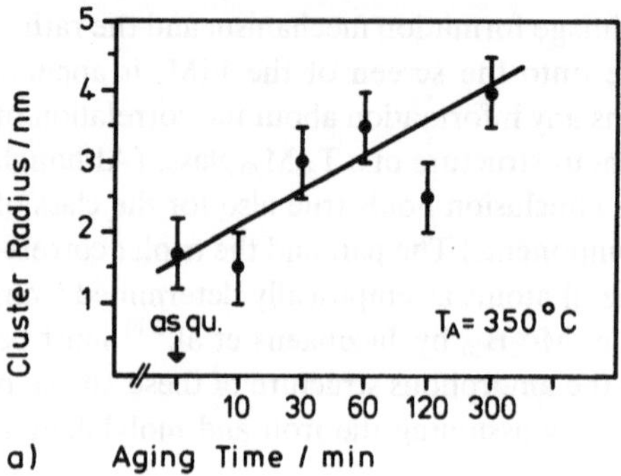

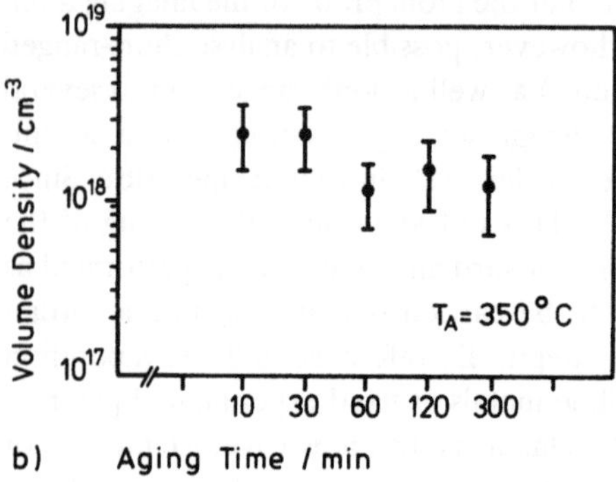

Fig. 4.33 a, b. Growth of the boron-rich clusters (**a**) in $Fe_{40}Ni_{40}B_{20}$ and their volume density (**b**) during isothermal aging at 350 °C. (After J. Piller[90])

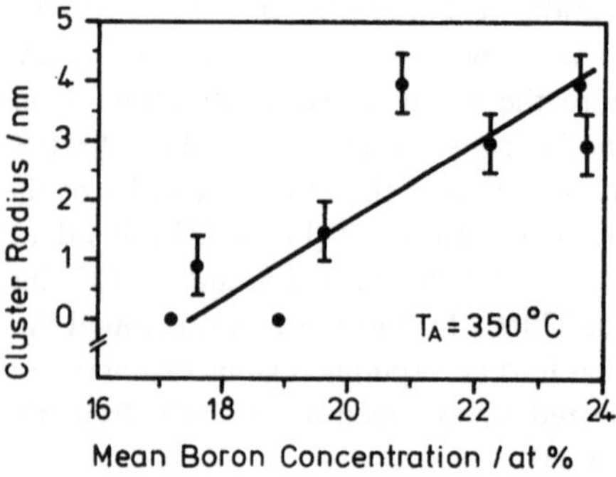

Fig. 4.34. Variation of the size of the Fe_3B clusters with the boron concentration of the surrounding amorphous matrix into which the clusters were embedded. (After J. Piller[90])

from the cluster size (e.g. the boron concentrations of the matrix and the clusters, and, hence, the precipitated volume and the number density of the precipitates) could be derived through a comparison of the experimentally obtained concentration profiles and autocorrelograms with the corresponding ones which were obtained from computer simulations. For this purpose composition profiles were generated on a computer in an analogous way to real atom-probe experiments from a $M_{80}B_{20}$ metallic glass, whose state of decomposition had been simulated with the above mentioned parameters. These parameters have been varied until the frequency distribution with respect to the height of the boron peaks in the concentration profiles, and the related autocorrelograms agreed with the corresponding experimental ones.

Earlier studies of TM metallic glasses by means of X-ray small-angle scattering, TEM and differential scanning calorimetry have suggested that some of these alloys are either already decomposed into two amorphous phases (e.g. Fe-B[267], Pd-Au-Si[268] and Ti-Zr-Be[269]) after liquid-quenching or reveal phase separation as a result of structural relaxation during annealing below the crystallization temperature[268]. Unlike these studies, in the atom-probe study of $Fe_{40}Ni_{40}B_{20}$ by Piller it became possible to follow the decomposition reaction quantitatively, especially with respect to distribution, spatial extension, and composition of the evolving amorphous phases, even though their compositional differences have been relatively small.

Attempts utilizing both X-ray energy-dispersive techniques and electron-energy loss spectroscopy to analyse the compositions of the multi-phase structures of amorphous Ti-Zr-Be alloys[269], which are already discernible in the TEM after liquid quenching, have not been successful. Judging from the current state of the art with respect to atom-probe analyses of metallic glasses, the compositional analyses of these rather complex phases in glassy Ti-Zr-Be ought to be amenable to this technique.

5 Conclusions and Outlook

The mechanisms of field evaporation and field-ion image formation are still not sufficiently well understood to predict the field evaporation characteristics and the contrast of an alloy, the microstructure and chemical composition of which are to be analysed by field-ion spectroscopy. This lack of a complete understanding of the basic mechanisms in fact does not severely limit the range of applicability of the field-ion techniques in materials science, but it often necessitates the imaging and field-evaporation characteristics to be determined empirically prior to the actual analysis; this empirical approach, however, can be rather tedious.

The range of applicability of a modern straight atom-probe FIM, which has been designed for metallurgical applications, appears currently not expansible by further instrumental improvements. The mass resolution of this instrument, which is inferior to that of the high-resolution energy-focusing atom probe, was found to be sufficient for all metallurgical investigations reported until today, and presumably might be further increased by replacing the high-voltage pulser by a laser pulser. The utilization of a high-resolution energy-focusing atom probe, the construction of which requires a larger instrumental expenditure, is of real advantage only for those studies where the field-evaporated ions suffer a large energy loss (e.g. in high purity semiconductors) or where single isotope resolution is routinely required; the latter is rarely realized in metallurgical investigations.

The imaging atom probe has not yet been employed to its full potential but must be considered to be particularly powerful in distinguishing phase heterogeneities which are not discernible in the FIM image. Considered as separate instruments, the atom-probe FIM appears to be markedly more versatile for metallurgical applications than the IAP. The feasible integration of an IAP into an atom-probe FIM, however, combines the advantages of both instruments; this fact will most likely be taken into account in future atom-probe designs.

Despite the fact that field-ion spectroscopy is a rather recent technique amongst the various microanalytical tools and considering that atom-probe FIMs are not commercially available, the technique has already been applied extensively to microstructural analyses of a great variety of crystalline metals and alloys, some metallic glasses, and also, in a first approach, to semiconductors. Due to its unique features such as high lateral and high depth resolution combined with simultaneous compositional analyses, the technique is seen to be most powerful for the investigation of point defects in metals, of composition profiles across interfaces, and of early stage decomposition in alloys, and, as a consequence, most comprehensive and systematic metallurgical studies have been reported – and most likely will be reported in the future – in these areas.

Acknowledgements. During writing this article I profited from many stimulating discussions with my colleague and friend Dr. P. A. Beaven to whom I express my sincere thanks. I owe my special thanks to Prof. P. Haasen for his continuous interest in our research work, for his encouragement when all the things that could go wrong did go wrong, his stimulating discussions, and for his critical reading of the manuscript. Many of the results presented originate from diploma and Ph. D. theses, carried out in our FIM group at the Institute for Metalphysics, University of Göttingen. I wish to thank the former members of this group, in particular Drs. K.-E. Biehl, J. Piller, H. Wendt, and Mr. U. Lorenz for their co-operation and for the provision of some unpublished results. The supply of original FIM micrographs by Dr. S. S. Brenner, Dr. A. J. Melmed, Prof. D. N. Seidman, Dr. A. R. Waugh and Prof. M. Yamamoto is also gratefully acknowledged. The Sonderforschungsbereich

126, Clausthal-Göttingen, provided the financial support to build and operate the atom-probe FIM. Mrs. I. Mann and Mr. L. Wilk kindly helped in preparing the manuscript.

Last, but certainly not least, I wish to thank my wife Margrit for her patience during all those years when the working day did not consist of just eight hours.

6 References

1. Müller, E. W.: Z. Physik *131*, 136 (1951)
2. Müller. E. W., Panitz, J. A.: 14th Field Emission Symp., NBS Washington, DC (1967)
3. Turner, P. J. et al.: J. Phys. (E), *2*, 731 (1969)
4. Waugh, A. R., Southon, M. J.: Surface Sci. *68*, 79 (1977)
5. Panitz, J. A.: J. Vac. Sci. Technol. *11*, 206 (1974)
6. Panitz, J. A.: Rev. Sci. Instrum. *44*, 1034 (1973)
7. Piller, J.: Diploma Thesis, University of Göttingen, 1977
8. Wagner, R.: Proc. 4th Tagung Mikrosonde, Dresden 1978, p. 173
9. Müller, E. W., Tsong, T. T.: Field Ion Microscopy, Principles and Applications, Elsevier Publ. Comp., New York 1969
10. Bowkett, K. M., Smith, D. A.: Field-Ion Microscopy, Defects in Crystalline Solids, Vol. 2 (Eds. Amelinckx, S., Gevers, R., Nihoul, J.) North Holland Publ. Comp., Amsterdam 1970
11. Melmed, A. J.: 26th Field Emission Symp., Berlin 1979
12. Edington, J. W.: Practical Electron Microscopy in Mat. Sci., McMillan Philips Technical Library, Eindhoven 1974
13. Loberg, B., Nordén, H.: Phil. Mag. *16*, 1147 (1967)
14. Waugh, A. R.: 25th Field Emission Symp., Albuquerque 1978
15. Piller, J.: Ph. D. Thesis, Univ. of Göttingen, 1981
16. Sakurai, T., Melmed, A. J.: 25th Field Emission Symp., Albuquerque 1978
17. Melmed, A. J., Carroll, J. J.: 24th Field Emission Symp., Oxford 1977
18. Brenner, S. S., Kinney, J. T.: Surf. Sci. *31*, 88 (1970)
19. Müller, E. W.: Advances in Electronics and Electron Physics, Vol. 13, Academic Press, New York 1960, p. 83
20. Chen, Y. C., Seidman, D. N.: Surf. Sci. *26*, 61 (1971)
21. Müller, E. W., Young, R. D.: J. Appl. Phys. *32*, 2425 (1961)
22. Tsong, T. T., Müller, E. W.: Phys. Rev. Letters *25*, 911 (1970)
23. Tsong, T. T., Müller, E. W.: phys. stat. sol. (a) *1*, 513 (1970)
24. Müller, E. W., McLane, S. B., Panitz, J. A.: Surf. Sci. *17*, 430 (1969)
25. Rendulic, K. D.: ibid. *28*, 285 (1971)
26. Waugh, A. R., Boyes, E. D., Southon, M. J.: ibid. *61*, 109 (1976)
27. Tsong, T. T., Müller, E. W.: J. Phys. *55*, 2884 (1971)
28. Müller, E. W.: Surface Sci. *8*, 462 (1967)
29. Bassett, D. W.: Surface and Defect Properties of Solids, Vol. 2, chapter 2, 34 (1973)
30. Müller, E. W.: Phys. Rev. *102*, 618 (1956)
31. Gomer, R., Swanson, L. W.: J. Chem. Phys. *38*, 1613 (1963)
32. Müller, E. W., Tsong, T. T.: Field Ion Microscopy, Field Ionization and Field Evaporation, in: Progress in Surface Science, Vol. 4, Part 1 (Ed. Davison, S.) Pergamon Press, 1973; Tsong, T. T.: Surf. Sci. *70*, 211 (1978)
33. Tsong, T. T.: J. Chem. Phys. *54*, 4205 (1971)
34. Ernst, N.: Surf. Sci. *87*, 469 (1979)
35. Brenner, S. S., McKinney, J. T.: Appl. Phys. Letters *13*, 29 (1968)
36. Taylor, D. M.: Ph. D. Thesis, Cambridge, 1970
37. Waugh, A. R., Boyes, E. D., Southon, M. J.: Surface Sci. *61*, 109 (1976)
38. Southon, M. J. et al.: ibid. *53*, 554 (1975)
39. Ehrlich, G., Kirk, C. F.: ibid. *48*, 1465 (1968)
40. Plummer, E. W., Rhodin, T. N.: ibid. *49*, 3479 (1968)
41. Boyes, E. D., Southon, M. J.: Vacuum *22*, 447 (1972)
42. Miller, M. K., Smith, G. D. W.: Metal Science *11*, 249 (1977)
43. Boyes, E. D. et al.: 22nd Field Emission Symp., Atlanta, 1975
44. Andrén, H. O., Nordén, H.: Scand. J. Metallurgy *8*, 147 (1979)
45. Wagner, R.: unpublished result
46. Turner, P. J.: Metal Science *9*, 370 (1975)
47. Goodmann, S. R., Brenner, S. S.: 18th Field Emission Symp., Eindhoven 1971
48. Turner, P. J., Southon, M. J.: 16th Field Emission Symp., Pittsburgh 1969

49. Eaton, H. C., Bayuzick, R. J.: Surf. Science *70*, 408 (1978)
50. Eaton, H. C., Gipson, G. S.: 26th Int. Field Emission Symp., Berlin 1979
51. Smith, P. J., Smith, D. A.: Phil. Mag. *21*, 907 (1970)
52. Birdsye, P. J., Smith, D. A.: Surf. Science *23*, 198 (1970)
53. Wendt, H.: Ph. D. Thesis, Univ. of Göttingen, 1981
54. Seidman, D. N.: J. Phys. F: Metal Physics *3*, 393 (1973)
55. Scalan, R. M., Styris, D. L., Seidman, D. N.: Phil. Mag. *23*, 1439 (1971)
56. Pétroff, P., Seidman, D. N.: Appl. Phys. Letters, *18*, 518 (1971)
57. Gold, E., Machlin, E. S.: Phil. Mag. *18*, 453 (1968)
58. Chen, C. G., Balluffi, R. W.: Acta Met. *23*, 911 (1975) and *23*, 931 (1975)
59. Page, T. F., Ralph, B.: Proc. Roy. Soc. Lond. *A 339*, 223 (1974)
60. Dubroff, W., Machlin, E. S.: Acta Met. *16*, 1313 (1968)
61. Southworth, H. N., Ralph, B.: J. Microsc. *90*, 167 (1970)
62. Page, T. F., Ralph, B.: Surf. Sci. *36*, 9 (1973)
63. Berger, A. S., Seidman, D. N., Balluffi, R. W.: Acta Met. *21*, 123 (1973)
64. Papazian, J. M.: J. Microsc. *95*, 429 (1972)
65. Tsong, T. T.: Surf. Sci. *10*, 303 (1968)
66. Newman, R. W., Hren, J. J.: Phil. Mag. *16*, 211 (1967)
67. Yamamoto, M. et al.: Japan. J. Appl. Phys. *13*, 1461 (1974)
68. Yamamoto, M. et al.: ibid. *14*, 703 (1975)
69. Taunt, R. J., Sinclair, R., Ralph, B.: phys. stat. sol. (a) *16*, 469 (1973)
70. Taunt, R. J., Ralph, B.: Surf. Sci. *47*, 569 (1975)
71. Taunt, R. J., Ralph, B.: phys. stat. sol. *29*, 431 (1975)
72. Brenner, S. S., Miller, M. K.: Proc. of 27th Int. Field Emission Symp. (Eds. Yashiro, Y., Igata, N.) Tokyo, 1980, p. 238
73. Youle, A. et al.: J. Microscopy *95*, 309 (1972)
74. Davies, D. M., Ralph, B.: ibid. *96*, 343 (1972)
75. Wendt, H.: Diploma thesis, Univ. of Göttingen, 1978.
    Wendt, H., Wagner, R.: Acta Met. (1982), in the press
76. Seidman, D. N., Lie, K. H.: Acta Met. *20*, 1045 (1972)
77. Seidman, D. N.: Surf. Sci. *70*, 532 (1978)
78. Wilkes, T. J., Smith, G. D. W., Smith, D. A.: Metallography *7*, 403 (1974)
79. Drechsler, M., Wolf, P.: Proc. IV. Int. Congress Electron Microscopy, Berlin 1958 (Springer-Verlag, Berlin 1960), Vol. 1, p. 835
80. Wagner, R., Brenner, S. S.: Acta Met. *26*, 197 (1978)
81. Brenner, S. S., Wagner, R., Spitznagel, J. A.: Met. Trans. *9 A*, 1961 (1978)
82. Chen, Y. D., Seidman, D. N.: Surf. Sci. *26*, 61 (1971)
83. Brenner, S. S., McKinney, J. T.: Surf. Sci. *23*, 88 (1970)
84. Miller, M. K., Beaven, P. A., Smith, G. D. W.: Surf. and Interface Analysis *1*, 149 (1979)
85. Panitz, J. A., Walko, R. J.: Sci. Instr. *47*, 1251 (1976)
86. Panitz, J. A.: Sandia Lab. Report SAND 75-0116 (1975)
87. Wagner, A., Hall, T. M., Seidman, D. N.: Rev. Sci. Instr. *46*, 1032 (1975)
88. Brenner, S. S., McKinney, J. T.: ibid. *43*, 1264 (1972)
89. Biehl, K. E., Wagner, R.: Proc. of 27th Int. Field Emission Symp., (Eds. Yashiro, Y., Igata, N.) Tokyo 1980, p. 267
90. Piller, J.: ibid. p. 285, and Ph. D. Thesis, Univ. of Göttingen, 1980
91. Wendt, H.: ibid. p. 280
92. Hill, S, A., Ralph, B.: Proc. Conf. on Phase Transformation, York, Vol. 2, Chameleon Press, London 1979, p. 44
93. Krishnaswamy, S. V., Lane, S. B., Müller, E. W.: J. Vac. Sci. Technol. *11*, 899 (1974)
94. Waugh, A. R.: J. Phys. E *11*, 49 (1978)
95. Müller, E. W., Krishnaswamy, S. V.: Rev. Sci. Instr. *46*, 1053 (1974)
96. Poschenrieder, W. P.: Int. J. Mass Spectrom. Ion Phys. *9*, 357 (1972)
97. Nishikawa, O. et al.: Proc. 27th Int. Field Emission Symp., (Eds. Yashiro, Y., Igata, N.) Tokyo 1980, p. 91
98. Walko, R. J., Müller, E. W.: phys. stat. sol. (a) *9*, K 9 (1972)
99. Panitz, J. A.: Progress Surf. Sci. *8*, 219 (1978)

100. Wagner, R.: Phys. Blätter *36*, 65 (1980)
101. Cockayne, D. J. H.: Z. Naturforschg. *27a*, 452 (1972)
102. Komoda, T.: Japan. J. Appl. Phys. *3*, 122 (1964)
103. Paris, D., Lesbats, P., Levy, J.: Scripta Met. *9*, 1373 (1975)
104. Müller, E. W.: Proc. Int. Symp. Reactivity of Solids, (Ed. de Boer et al.), Elsevier, Amsterdam 1960, p. 691
105. Fortes, M. A., Ralph, B.: Acta Met. *15*, 707 (1967)
106. Machlin, E. S.: Trans. A. S. M. *60*, 260 (1967)
107. Papazian, J. P.: J. Microscopy *95*, 429 (1972)
108. Gold, E., Machlin, E. S.: Phil. Mag. *18*, 453 (1968)
109. Machlin, E. S.: ibid. *18*, 465 (1968)
110. Machlin, E. S.: Scripta Met. *10*, 521 (1976)
111. Clapp, P. C., Moss, S. C.: Phys. Rev. *142*, 418 (1966)
112. Galligan, J. M.: Vacancies and Interstitials in Metals (Eds. Seeger, A. et al.) North-Holland, Amsterdam 1970, p. 557
113. Müller, E. W.: ibid., p. 575
114. Seidman, D. N., Wilson, K. L., Nielsen, C. H.: Fundamental Aspects of Radiation Damage in Metals (Eds. Robinson, M. T., Young, Jr. F. W.) Nat. Techn. Information Service, U.S. Dept of Commerce, Springfield 1975, p. 373
115. Seidman, D. N.: Radiation Damage in Metals (Eds. Peterson, N. L., Harkness, S. D.) Amer. Soc. Metals, Metals Park 1976, p. 28
116. Seidman, D. N.: Surf. Sci. *70*, 532 (1978)
117. Wei, C. Y., Seidman, D. N.: Appl. Phys. Letters *34*, 622 (1979)
118. Igata, N., Sato, S., Shibata, K.: Proc. 27th Int. Field Emission Symp., (Eds. Yashiro, Y., Igata, N.) Tokyo 1980, p. 233
119. Stiller, K., Nordén, H.: Proc. 27th Int. Field Emission Symp., (Eds. Yashiro, Y., Igata, N.) Tokyo 1980, p. 209
120. Wilson, K. L., Seidman, D. N.: Rad. Effects *33*, 149 (1977)
121. Scanlan, R. M., Styris, D. L., Seidman, D. N.: Phil. Mag. *23*, 1459 (1971)
122. Wilson, K. L., Baskes, M. I., Seidman, D. N.: Acta Met. *28*, 89 (1980)
123. Wei, C. Y., Seidman, D. N.: Phil. Mag. *A 37*, 257 (1978), and Rad. Effects *32*, 229 (1977)
124. Robertson, A. H., Seidman, D. N.: J. Phys. *E* (Sci. Instr.) *1*, 1244 (1968)
125. Scanlan, R. M. et al.: Cornell Mat. Science Center Rep. No. 1159 (1969)
126. Robinson, M. T., Torrens, I. M.: Phys. Rev. *B 9*, 5008 (1974)
127. Attardo, M. J., Galligan, J. M.: Phys. Rev. Letters *14*, 671 (1965)
128. Bowkett, K. M., Ralph, B.: Proc. Roy. Soc. *A 312*, 51 (1969)
129. Brenner, S. S., Seidman, D. N.: Radiation Effect *14*, 73 (1975)
130. Jeanotte, D., Galligan, J. M.: Acta Met. *18*, 71 (1970)
131. Wagner, A., Seidman, D. N.: J. Nucl. Mat. *83*, 48 (1979)
132. Seeger, A.: Fundamental Aspects of Radiation Damage in Metals, (Eds. Robinsen, M. T., Young, F. W.) Nat. Techn. Information Service, U.S. Dpt. of Commerce, Springfield, Virginia 1975, p. 493
133. Frank, W., Seeger, A.: Rad. Effects *25*, 17 (1975)
134. Pétroff, P., Seidman, D. N.: Acta Met. *21*, 323 (1973)
135. Kim, Y.-W., Galligan, J. M.: ibid *26*, 379 (1978)
136. Wei, C. Y., Seidman, D. N.: J. Nucl. Mat. *69 & 70*, 693 (1978)
137. Attardo, M. J., Galligan, J. M.: Phys. Rev. Lett. *17*, 1173 (1966)
138. Attardo, M. J., Galligan, J. M., Chow, J. G. Y.: Phys. Rev. Lett. *19*, 73 (1967)
139. Seidman, D. N.: Scripta Met. *13*, 251 (1979)
140. Wilson, K. L., Seidman, D. N.: Radiation Effects *27*, 67 (1975)
141. Panitz, J. A.: J. Sci. Technol. *14*, 502 (1977)
142. Chandrasekharaiah, M. N., Ranganathan, S.: phys. stat. sol. (a) *13*, 259 (1972)
143. Taunt, R. J., Sinclair, R., Ralph, B.: phys. stat. sol. (a) *16*, 469 (1973)
144. Yamamoto, M. et al.: Japan. J. Appl. Phys. *11*, 437 (1972)
145. Yamamoto, M. et al.: J. Phys. Soc. Japan *36*, 1330 (1974)
146. Yamamoto, M. et al.: Japan. J. Appl. Phys. *13*, 1461 (1974)

147. Yamamoto, M. et al.: Proc. Int. Symp. on Applic. of FIM to Metallurgy (Eds. Hasiguti, R. R., Yashiro, Y., Igata, N.) Lake-Yamanaka 1976, p. 112
148. Yamamoto, M., Nenno, S., Shono, F.: Japan. J. Appl. Phys. *16*, 2045 (1977)
149. Tsong, T. T., Müller, E. W.: J. Appl. Phys. *38*, 3531 (1967)
150. Sinclair, R., Ralph, B., Leake, J. A.: Phil. Mag. *28*, 1111 (1973)
151. Sinclair, R., Leake, J. A., Ralph, B.: phys. stat. sol. (a) *26*, 285 (1974)
152. Taunt, R. J., Ralph, B.: ibid. *24*, 207 (1974)
153. Taunt, R. J., Ralph, B.: Surf. Sci. *47*, 569 (1975), and phys. stat. sol. (a) *29*, 431 (1975)
154. Southworth, H. N., Ralph, B.: Phil. Mag. *14*, 383 (1966)
155. Ruedl, E., Delavignette, P., Amelinckx, S.: phys. stat. sol. (a) *28*, 305 (1968)
156. Lefevre, B. G., Grenga, H., Ralph, B.: Phil. Mag. *18*, 1127 (1968)
157. Lefevre, B. G., Newman, R. W.: Applications of FIM in Physical Metallurgy and Corrosion. Georgia Inst. of Technol., Atlanta, Ga. 1969, p. 243
158. Ranganathan, S., Lyon, H. B., Thomas, G.: J. Appl. Phys. *38*, 4957 (1967)
159. Chakravarti, B., Starke, E. A., Lefevre, B. G.: J. Met. Sci. *5*, 394 (1970)
160. Newman, R. W., Lefevre, B. G.: Phil. Mag. *19*, 241 (1969)
161. Taunt, R. J., Ralph, B.: ibid *30*, 1379 (1974)
162. Taunt, R. J.: Ph. D. Thesis, Cambridge Univ., 1973
163. Brandon, D. G. et al.: Acta Met. *12*, 813 (1964)
164. Bishop, G. H., Chalmers, B.: Scripta Met. *2*, 133 (1968)
165. Bollmann, W.: Crystal Defects and Crystalline Interfaces, Springer-Verlag, Berlin 1970
166. Warrington, D. W., Grimmer, H.: Phil. Mag. *30*, 461 (1974)
167. Gleiter, H., Chalmers, B.: Prog. Mat. Sci. *16* (1972)
168. Weins, M., Weins, J. J.: Phil. Mag. *26*, 885 (1972)
169. Smith, D. A., Pond, R. C., Vitek, V.: Acta Met. *25*, 475 (1977)
170. Smith, D. A.: Can. Metall. Q. *30*, 65 (1975)
171. Loberg, B., Norden, H.: Grain Boundary Structure and Properties, (Eds. Chadwick, G. A., Smith, D. A.) Academic Press, London 1976, p. 1
172. Weins, M., Gleiter, H., Chalmers, B.: J. Appl. Phys. *42*, 2639 (1971)
173. Loberg, B., Norden, H., Smith, D. A.: Phil. Mag. *24*, 897 (1971)
174. Smith, D. A. et al.: ibid. *17*, 1065 (1968)
175. Smith, D. A., Smith, G. D. W.: Phys. Bull. *21*, 393 (1970)
176. Howell, P. R., Page, T. F., Ralph, B.: Phil. Mag. *25*, 879 (1972)
177. Smith, D. A., Goringe, M. J.: Phil. Mag. *25*, 1505 (1972)
178. Fortes, M. A.: Surf. Sci. *28*, 95 (1971)
179. Fortes, M. A.: Surf. Sci. *28*, 117 (1971)
180. Fortes, M. A., Smith, D. A.: J. Appl. Phys. *41*, 2348 (1970)
181. Page, T. F., Howell, P. R., Ralph, B.: ibid. *44*, 902 (1973)
182. Beaven, P. A., Smith, D. A., Miller, M. K.: Phil. Mag. A *43*, 1063 (1981)
183. Ranganathan. S.: J. Appl. Phys. *37*, 4346 (1966)
184. Clark, W. A. T., Smith, D. A., Suryanarayana, C.: Can. Met. Q. *13*, 49 (1974)
185. Smith, D. A.: Scripta Met. *13*, 379 (1979)
186. Ishida, Y., Smith, D. A.: ibid. *8*, 293 (1974)
187. Loberg, B., Smith, D. A.: Ark. Fys. *40*, 513 (1970)
188. Pond, R. C., Smith, D. A.: Can. Met. Q. *13*, 39 (1974)
189. Clark, W. A. T., Smith, D. A.: Phil. Mag. A *38*, 367 (1978)
190. Pumphrey, P. H.: Grain Boundary Structure and Properties, (Eds. Chadwick, G. A., Smith, D. A.) Academic Press, London 1976, p. 139
191. Warrington, D. W.: Inst. Metallurgists Review Course on Grain Boundary Structure, Series 2, No. 7, 407-71-Y
192. Morgan, R., Ralph, B.: Acta Met. *15*, 341 (1967)
193. Bayuzick, R. J., Goedrich, R. S.: Surf. Sci. *23*, 225 (1973)
194. Bolin, P. L., Bayuzick, R. J., Ranganathan, B. N.: Phil. Mag. *31*, 891 (1975)
195. Bolin, P. L., Bayuzick, R. J., Ranganathan, B. N.: ibid. *31*, 895 (1975)
196. Howell, P. R., Ralph, B.: J. Microsc. *102*, 361 (1974)
197. Hildon, A. et al.: ibid. *99*, 29 (1973)
198. Hildon, A. et al.: ibid. *99*, 41 (1973)

References

199. Wilkes, T. J. et al.: Proc. Int. Symp. on Appl. of FIM to Metallurgy (Eds. Hasiguti, R. R., Yashiro, Y., Igata, N.) Lake Yamanaka 1976, p. 38
200. Faulkner, R. G., Ralph, B.: Acta Met. *20*, 703 (1972)
201. Aaronson, H. I.: J. Microsc. *102*, 275 (1974)
202. Ryan, H. F., Sniter, J.: Phil. Mag. *10*, 727 (1964)
203. Fortes, M. A., Ralph, B.: Acta Met. *15*, 707 (1967)
204. Smith, D. A., Smith, G. D. W.: 3rd Int. Conf. Strength of Metals, Cambridge 1973
205. Turner, P. J., Papazian, J. M.: Metal Sci. *7*, 81 (1973)
206. Bach, P. W., Beyer, J., Verbraak, C. A.: Scripta Met. *14*, 205 (1980)
207. Bach, P. W., Verbraak, C. A.: Proc. 27th Int. Field Emission Symp., (Eds. Yashiro, Y., Igata, N.) Tokyo 1980, p. 254
208. Miller, M. K., Beaven, P. A., Smith, G. D.W.: Proc. Conf. Phase Transformations – York 2 (1979), Institution of Metallurgists, Conf. Ser. 3 No. 11, London, p. II-114
209. Ng, Y. D., Tsong, T. T., McLane, S. B.: Surf. Sci. *84*, 31 (1978)
210. Mintz, B., Turner, P. J.: Met. Trans. *9A*, 1611 (1978)
211. Brenner, S. S., Miller, M. K.: Proc. 27th Int. Field Emission Symp., (Eds. Yashiro, Y., Igata, N.) Tokyo 1980, p. 238
212. Williams, P. R. et al.: Proc. Conf. Phase Transformations – York 2 (1979), Institution of Metallurgists, Conf. Ser. 3 No. 11, London, p. II-98
213. Miller, M. K., Smith. G. D. W.: Met. Sci. *11*, 249 (1977)
214. Miller, M. K. et al.: Surf. Sci. *70*, 470 (1978)
215. Kinsman, K. R., Aaronson, H. I.: Transformation and Hardenability in Steels, Climax Molybdenum, Ann Arbor 1967, p. 33
216. Brenner, S. S., Myers, S. M.: Proc. 27th Int. Field Emission Symp., (Eds. Yashiro, Y., Igata, N.) Tokyo 1980, p. 219
217. Waugh, A. R., Paetke, S., Edmonds, D. V.: Met. Sci. 1981, in press
218. Hilliard, J. E.: Phase Transformation, chapter 12, Amer. Soc. Metals, Metals Park, Ohio 1970
219. Cahn, J. W.: Trans. AIME *242*, 166 (1968), and Acta Met. *14*, 1685 (1966)
220. Langer, J. S.: Fluctuations, Instabilities and Phase Transformations (Ed. Riste, T.) p. 19, Plenum Press, N.Y. 1975
221. Goodman, S. R., Brenner, S. S., Low, J. R.: Met. Trans *4*, 2363 (1973) and *4*, 2371 (1973)
222. Youle, A., Ralph, B.: Met. Sci. J. *6*, 149 (1972)
223. Schwartz, D. M., Ralph, B.: Phil. Mag. *19*, 1061 (1969)
224. Wagner, C.: Elektrochem. *65*, 681 (1961)
225. Ardell, A. J., Nicholson, R. B.: Acta Met. *14*, 1295 (1966)
226. Hill, S. A., Ralph, B.: Proc. Conf. Phase Transformation – York (1979), Vol. 2, Institution of Metallurgists Conf. Ser. 3 No. 11, London. p. 44
227. Beaven, P. A., Miller, M. K., Smith, G. D. W.: Proc. Inst. of Phys. Conf. Ser. *36*, 199 (1977)
228. Beaven, P. A. et al.: Proc. 9th Int. Congr. on Electron Microscopy, Vol. 1, Toronto 1978, p. 626
229. Schwartz, D. M., Davenport, A. T., Ralph, B.: Phil. Mag. *18*, 431 (1968)
230. Schwartz, D. M., Ralph, B.: Phil. Mag. *19*, 1069 (1969)
231. Brenner, S. S., Goodman, S. R.: Scripta Met. *5*, 865 (1971)
232. Driver, J. H., Papazian, J. M.: Acta Met *21*, 1139 (1973)
233. Turner, P. J.: Proc. Int. Symp. on Appl. of FIM to Metallurgy, (Eds. Hasiguti, R. R., Yashiro, Y., Igata, N.) Lake Yamanaka 1976, p. 69
234. Turner, P. J., Papazian, J. M.: Met. Sci. *7*, 81 (1973)
235. Andrén, H.-O., Henjered, A., Nordén, H.: J. Mat. *15*, 2365 (1980)
236. Huffman, G. P., Podgurski, H. H.: Acta Met. *23*, 1367 (1975)
237. Jack, K. H.: Treatment, I.S.I., London 1973, p. 39
238. Martin, J. W., Doherty, R. D.: Stability of Microstructure in Metallic Systems, Cambridge Univ. Press, Cambridge 1976
239. Youle, A., Schwartz, D. M., Ralph, B.: Met. Sci. J. *5*, 131 (1971)
240. Aaron, H. B., Fainstain, D., Kotler, G. R.: J. Appl. Phys. *41*, 4404 (1970)
241. Abe, T., Hirano. K.: 26th Int. Field Emission Symp., Berlin 1979, p. 54
242. Butler, E. P., Thomas, G.: Acta Met. *18*, 347 (1970)
243. Livak, R. J., Thomas, G.: Acta Met. *19*, 497 (1971)

114

244. Watts, A. J.: Ph. D. Thesis, Cambridge Univ., 1975
245. Saito, K., Watanabe, R.: Japan. J. Appl. Phys. *8*, 14 (1969)
246. Ben Israel, D. H., Fine, M. E.: Acta Met. *11*, 1051 (1963)
247. Watts, A. J., Ralph, B.: Acta Met. *25*, 1013 (1977)
248. Laughlin, D. E., Sinclair, R., Tanner, L. W.: Scripta Met. *14*, 373 (1980)
249. Laughlin, D. E., Cahn, J. W.: Acta Met. *23*, 329 (1975)
250. Tsujimoto, T.: Trans. Jap. Inst. Met. *18*, 393 (1977)
251. Biehl, K. E.: Ph. D. Thesis, Univ. of Göttingen, 1980
252. Miyazaki, T.: Trans. JIM *12*, 119 (1972)
253. Watts, A. J., Ralph, B.: Surf. Sci. *70*, 459 (1978)
254. Wagner, R.: Czech. J. Phys. *B 21*, 198 (1981)
255. Melmed, A. J., Stein, R. J.: Surf. Sci. *49*, 645 (1975)
256. Ernst, L., Block, J. H.: Surf. Sci. *49*, 293 (1975)
257. Tsong, T. T., Ng, Y. S., Melmed, A. J.: Surf. Sci. *77*, L 187 (1978)
258. Ohno, Y. et al.: Surf. Sci. *69*, 521 (1977)
259. Yamamoto, M., Seidman, D. N.: 27th Int. Field Emission Symp., (Eds. Yashiro, Y., Igata, N.) Tokyo 1980, p. 317
260. Sakurai, T., Culbertson, R. J., Melmed, A. J.: Surf. Sci. *78*, L 221 (1978)
261. Melmed, A. J. et al.: 27th Int. Field Emission Symp. (Eds. Yashiro, Y., Igata, N.) Tokyo 1980, p. 430
262. Melmed, A. J., Carroll, J. J., Givargizov, G. I.: Proc. 26th Int. Field Emission Symp., Berlin 1979, p. 70
263. Kellogg, G. L., Tsong, T. T.: ibid. p. 78
264. Brenner, S. S., Wagner, R.: 22nd Int. Field Emission Symp., Atlanta, Ga. 1975
265. Inal, I. T., Keller, L., Yost, F. G.: J. Mat. Sci. *15*, 1947 (1980)
266. Jacobaeus, P. et al.: Phil. Mag. *B 41*, 11 (1980)
267. Walter, J. L., Bartram, S. F.: Rapidly Quenched Metals III, Vol. 1 (Ed. Cantor, B.) The Metals Soc., London 1978
268. Chou, C.-P., Turnbull, D.: J. Non-Cryst. Sol. *17*, 169 (1975)
269. Tanner, L. E., Ray, R.: Scripta Met. *14*, 657 (1980)

Author Index Volumes 1–6

Structure and Bonding

Editors: M. J. Clarke, J. B. Goodenough,
P. Hemmerich (†), J. A. Ibers, C. K. Jørgensen,
J. B. Neilands, D. Reinen, R. Weiss,
R. J. P. Williams

Springer-Verlag
Berlin
Heidelberg
New York